To John

Best wishes

Matt Kilroel

Becca

The Changing World of Farming in Brexit UK

The 2016 referendum resulted in a vote for the United Kingdom to withdraw from the European Union. This has led to frenzied political debate across the whole spectrum of policy, and agriculture is no exception.

For the first time in a generation, the future of agriculture is unclear and unfettered by the constraints and incrementalism of the Common Agricultural Policy. This book makes an empirical contribution to the Brexit debate, bringing a social dimension to agri-Brexit and sustainable agriculture discourses. Understanding the *social* in the context of farmers is vital to developing a way forward on food security and agricultural sustainability. Farmers are the recipients of the market and policy signals that link to global uncertainties and challenges. This book is a commitment to understanding farmers as occupiers and managers of land. Chapters in this book explore farmers' own aspirations and knowledge about patterns of land use and production, which underpin discussions around the environment and sustainability.

There is a deficit in understanding what kind of agricultural industry we now have, following years of restructuring and repositioning. This book is an attempt to address that deficit and will appeal to students and researchers exploring agriculture, food politics and rural sociology.

Matt Lobley is a rural social scientist drawing primarily on the disciplines of geography and rural sociology. His research focuses on understanding influences on and impacts of farm household behaviour. His main interests relate to the role of farm households in the management of the countryside and the environmental and social impacts of agricultural restructuring. One of his main areas of expertise is family life cycle and succession issues on family farms.

Michael Winter is a social scientist with disciplinary roots in sociology and political science and has an avid interest in interdisciplinary research and engagement with natural science, especially agricultural science and ecology. His current research interests are in food security, the sustainable intensification of agriculture, land tenure and occupancy, the historical sociology of farming, the history of agricultural policy, and agri-environmental issues.

Rebecca Wheeler is a social scientist with a broad background in rural geography, environmental sustainability and social anthropology. She has worked extensively with farmers and the wider rural community on a range of agricultural and environmental issues. Her specific research interests include farmer behaviour and well-being, interactions between agriculture, rural communities and the environment, and the cultural heritage of place and landscape.

Perspectives on Rural Policy and Planning
Series Editors: Andrew Gilg and Mark Lapping

This well-established series offers a forum for the discussion and debate of the often-conflicting needs of rural communities and how best they might be served. Offering a range of high-quality research monographs and edited volumes, the titles in the series explore topics directly related to planning strategy and the implementation of policy in the countryside. Global in scope, contributions include theoretical treatments as well as empirical studies from around the world and tackle issues such as rural development, agriculture, governance, age and gender.

Rural Change in Australia
Population, Economy, Environment
Edited by Rae Dufty-Jones and John Connell

European Integration and Rural Development
Actors, Institutions and Power
Michael Kull

Evaluating the European Approach to Rural Development
Grass-roots Experiences of the LEADER Programme
Edited by Leo Granberg, Kjell Andersson and Imre Kovách

Globalization and Europe's Rural Regions
Edited by John McDonagh, Birte Nienaber and Michael Woods

Service Provision and Rural Sustainability
Infrastructure and Innovation
Edited by Greg Halseth, Sean Markey and Laura Ryser

The Changing World of Farming in Brexit UK
Matt Lobley, Michael Winter and Rebecca Wheeler

For more information about this series, please visit: www.routledge.com/Perspectives-on-Rural-Policy-and-Planning/book-series/ASHSER-1035

The Changing World of Farming in Brexit UK

Matt Lobley, Michael Winter and Rebecca Wheeler

Routledge
Taylor & Francis Group
LONDON AND NEW YORK

First published 2019
by Routledge
2 Park Square, Milton Park, Abingdon, Oxon OX14 4RN

and by Routledge
52 Vanderbilt Avenue, New York, NY 10017

Routledge is an imprint of the Taylor & Francis Group, an informa business

British Library Cataloguing-in-Publication Data
A catalogue record for this book is available from the British Library

Library of Congress Cataloging-in-Publication Data
A catalog record has been requested for this book

ISBN: 978-1-4094-0971-7 (hbk)
ISBN: 978-1-315-61451-9 (ebk)

Typeset in Times New Roman
by Out of House Publishing

FSC
www.fsc.org
MIX
Paper from responsible sources
FSC® C013056

Printed and bound in Great Britain by
TJ International Ltd, Padstow, Cornwall

We wish to dedicate this book to the memory of Griff Davies. Griff worked as a farm business economist in the Agricultural Economics Unit, the forerunner to the Centre for Rural Policy Research, at Exeter in the 1960s, 1970s and 1980s. When he died in 2005 he left money to the University for a programme of research on agriculture in the South West with an emphasis on the well-being of farmers. We hope this book is a fitting tribute to his generosity and vision.

Contents

Figures

Tables

Boxes

Acknowledgements

This book is the culmination of several years of work on a number of research projects undertaken by the authors within the Centre for Rural Policy Research (CRPR) at the University of Exeter. Consequently, we have a number of people and organisations to thank for making it possible. Not least, the funders of our research: Devon County Council for supporting the Sustainable Rural Futures Programme (SRFP) which, among other things, funded the South West Farm Surveys; Defra for the Sustainable Intensification Research Platform (SIP) and other Defra-funded research we draw on; a legacy from Griff Davies which also supported the SW Farm Survey work; and the Prince's Countryside Fund for the detailed research on small family farms. Particular thanks go to the members of the SRFP advisory group: Phil Norrey, Stewart Barr, Kevin Bishop, Don Gobbett, Debbie Myhill, Nigel Stone and Steve Turner; also our gratitude to erstwhile CRPR colleagues Allan Butler and Hannah Chiswell for their work on the SW Farm Surveys, Rob Fish for his work on SRFP and Gavin Hugget for managing SIP. We are of course also indebted to the plethora of scientists, researchers, farmers, industry experts and other participants who were involved in planning, conducting, creating and analysing the findings from these studies. Also thanks to Hannah Wittman and the occupants of Room 170 at the Centre for Sustainable Food Systems at the University of British Columbia where Michael Winter worked on this book in the spring of 2018. We are grateful also to Ruth Anderson at Taylor & Francis for her patience and support in keeping the book on track and ensuring its completion.

None of the above is responsible for any errors, omissions and inadequacies in this book.

Abbreviations

AES	agri-environmental scheme
AFI	Agri-environmental Footprint Index
AHDB	Agriculture and Horticulture Development Board
APS	Annual Population Survey
bTB	bovine tuberculosis
BPS	Basic Payment Scheme
CAP	Common Agricultural Policy
CRPR	Centre for Rural Policy Research
CS	Countryside Stewardship
CSA	Community Supported Agriculture
Defra	Department for Environment, Food and Rural Affairs
ELS	Entry Level Stewardship
ESA	Environmentally Sensitive Area
ESU	European Size Unit
FAO	Food and Agriculture Organization
FBI	farm business income
FBS	Farm Business Survey
FFA	Family Farmers' Association
FMD	foot and mouth disease
FTE	full-time equivalent
FWAG	Farming and Wildlife Advisory Group
GCA	Groceries Code Adjudicator
GDP	Gross Domestic Product
GM	genetic modification
HGCA	Home-Grown Cereals Authority
HLS	higher level stewardship
HNV	High Nature Value
IFM	Integrated Farm Management
IFS	Integrated Farming System
IPCC	Intergovernmental Panel on Climate Change
LFA	less favoured area
MMB	Milk Marketing Board
NCD	non-communicable disease

NFI	net farm business income
NFU	National Farmers' Union
NVZ	Nitrate Vulnerable Zone
PCF	Prince's Countryside Fund
PMB	Potato Marketing Board
ROCE	return on capital employed
ROTCE	return on tenant's capital employed
RPM	Resale Price Maintenance
SFP	Single Farm Payment
SGM	Standard Gross Margin
SI	Sustainable Intensification
SIP	Sustainable Intensification Research Platform
SLR	Standard Labour Requirement
SMR	standardised mortality ratio
SO	Standard Output
SRFP	Sustainable Rural Futures Programme
SW	South West
UNESCO	United Nations Educational, Scientific and Cultural Organization
WCED	World Commission on Environment and Development
YEN	Yield Enhancement Network

1 Introduction

Farmers and Brexit: setting the scene

The referendum in June 2016, which resulted in a vote in favour of the withdrawal of the United Kingdom from the European Union, has led to frenzied political debate across the whole spectrum of policy – and agriculture is no exception. For the first time in a generation there is a debate on the future of agriculture which is unfettered by the constraints and incrementalism of the Common Agricultural Policy (CAP). Consequently, Brexit has spawned a plethora of reports and campaigns on how policy should be reformed and agriculture re-envisioned in a post-Brexit world. At the time of writing in March 2018, the UK government had just issued a command paper on the future of agriculture alongside an evidence compendium and a summary of stakeholder ideas with a deadline for consultation responses of the 8 May 2018 (Cm 9577, 2018). An Agriculture Bill is expected to be introduced to Parliament before the summer recess. It is clear from even a superficial look at these sources that the old certainties based on CAP-induced compromises between farming and environmental interests have splintered. There are calls to focus policy primarily on the provision of public goods through environmental outcomes (Helm, 2016) and other calls to remove regulatory red tape and free up farming to play its role within an economic growth agenda (NFU, 2016). Others are using the opportunity to turn the spotlight on human health and nutrition (Lang et al., 2017). What is also clear is that this seemingly unending stream of reports does not necessarily reflect a deepening and widening of the evidential base for policy reflection. Brexit debate on agriculture has not, in the main, been characterised by fresh empirical research on the nature of UK agriculture in the second decade of the twenty-first century. Consequently, many of the campaigning reports are based on assumptions about farming and policy that may or may not be rooted in contemporary realities. For example, there is tendency to present farmers as a relatively homogeneous group at best segmented by the traditional categorisations of size, enterprise or tenure. Yet, as we show in Chapter 3 of this book, none of these familiar means of classifying farmers quite works in current circumstances and some oft-used data sources can be misleading. So one of the aims of this book is to make an empirical contribution to the Brexit debate about agriculture. We will say a little more about our data sources in the final section of this opening chapter.

A second observation we make about the great agri-Brexit debate is the low profile given to farmers, qua farmers. The majority of policy proponents are seeking policy measures that will influence farmers and farming to serve wider ends, such as through the provision of public goods (nature conservation, public access to the countryside etc.), or improved diet-related health outcomes. Even those campaigning on behalf of farmers are inclined to emphasise the contribution agriculture can make to UK PLC through exports and economic growth. And yet the provision of both market *and* public goods requires 'providers', in this case people we usually call farmers, although 'land managers and workers' might be a better term as not all rural land is necessarily worked or managed by conventionally defined farmers. Who actually owns, runs and works our farmed land is an empirical question. And a rather important one, even if we are concerned only with influencing market and public good outputs from the land – we surely need to know who, where and how to target these new policies. Moreover, we would argue that farmers are intrinsically significant in their own right. Like any other occupational group, farming comprises a wide range of individuals with different needs, behavioural norms, values and so forth. As a group, even in late modernity, they remain culturally distinctive.

While the policy and academic communities may struggle to 'place' farmers within their various discourses, there is no such ambivalence within popular culture as evidenced by TV programmes such as *Country File* and well-selling literary accounts of farming and rural life by writers such as Rebanks (2015). Here, farmers are distinctive and interesting, their survival and well-being to be valued. There is an academic language that offers some potentiality here, which is the notion of sustainability as encompassing economic, environmental *and social* elements. Our commitment to finding a way of bringing the social into agri-Brexit and sustainable agriculture discourses is one of the chief motivations for writing this book. Later in this chapter we make an attempt to explain the neglect of the social in so much of contemporary discussions of agriculture.

Another key theme of this book is the fitness of purpose of contemporary farming given the radically changing demands it now faces. The quest for sustainable farming that has dominated the policy agenda of developed countries for the past two decades, with its language of multifunctionality and environmental protection, is now confronted by the challenges of food, energy and water security within the context of climate change, as well as the profound policy challenges and uncertainties of Brexit. At no time since the Second World War has agriculture been so central to the foremost policy challenges confronting society. This is true globally for the challenges are global; agricultural commodities are increasingly traded internationally and the food price spikes of 2008 had economic, social and political effects across the world. But it is a grave mistake to assume that any global economic sector loses its spatiality, its particularity, its context. More than most economic activities, agriculture is rooted in locality. As any farmer will confirm, no two farms are the same. Soil, climate, topography, ecology are physical attributes that combine to deliver distinctiveness, and this distinctiveness is reinforced by the social and economic heterogeneity of an industry that encompasses everything from the small semi-subsistence holding to massive agri-business. Moreover, farming and the land have cultural and political resonance contributing to contrasting national and regional identities and understanding. These identities are often deeply conflicted, as exemplified in the response from one commentator to the designation of the Lake District as a UNESCO World Heritage Site. Describing the designation as a 'betrayal of the living world', Monbiot (2017) regards the new status as antithetical to conservation objectives in its preservation of sheep farming as a cultural practice. He argues that the portrayal of sheep farming in the Lake District as 'wholly authentic' and 'harmonious' with wildlife is part of a powerful cultural myth that obscures the widespread ecological damage associated with the practice. In Monbiot's view, public money spent on subsidising sheep farmers would be better spent on ecological restoration. Such debates draw into focus the contested nature of land and the question of how (agri)cultural landscapes are valued alongside preferences for ecological wilderness (and indeed other interests such as carbon storage).

Our contribution to the global issue of sustainable agriculture is to offer particularity. The global is made up of distinctive, particular places and we eschew any approach that loses all-important detail in the pursuit of generality to the detriment of focus and of purpose. So our focus on England, with a bias towards the south-west of England, should require no justification. This book is an attempt to understand some key issues facing agriculture in one part of the world, a particular and specific manifestation of global issues in a local context, and by so doing we hope to stimulate thought on how things are the same or differ in other places.

That our aspirations are in no way narrow or parochial is, hopefully, supported by our intention to bridge gaps and make connections. We seek to do so across a number of levels but three urgent imperatives are of particular importance. First, there is the *academic* imperative to bridge gaps between disciplines; in this instance to bring the insights of a range of social sciences into connection with environmental and agricultural sciences. Our underlying approach is rooted in three social sciences – human geography, political science, sociology – but our intention is both to learn lessons from other disciplines and to understand the technological and agro-ecological issues that underlie the industry we are interested in. Second, there is the *policy* imperative to connect sustainability and agricultural production, a concern epitomised in the mantra of 'sustainable intensification'. Third, we believe there is an *ethical* imperative to (re)connect farmers, to each other, to the geographical communities in which they live, and to the wider economy and society.

At the heart of our book is a commitment to understanding farmers, as occupiers and managers of land, for it is farmers who are the recipients of the market and policy signals that derive from global uncertainties and challenges. Their behavioural response and indeed their own aspirations and knowledge give rise to patterns of land use and production, the canvas of so many discussions and wider societal ambition around ecosystem services, natural capital and sustainability. In short, social and economic drivers have great bearing on the science of sustainability. And yet there is a deficit of understanding of what kind of agricultural industry we now have, following years of restructuring and repositioning. There are a number of factors that have contributed to this neglect. One of them, which we look at in more detail in Chapter 3, is the declining analytical and explanatory power of the annual agricultural census/survey. This longstanding exercise in data gathering has failed to keep pace with the growing complexity and heterogeneity of agricultural occupancy and business arrangements. Another is the lamentable decline and fragmentation of social scientific interest in agriculture, which we seek to remedy throughout the book. As a society, we may have a reasonable knowledge of the aggregate economics of the industry but what about its people? Who are the farmers now and is the agricultural industry socially and culturally equipped for the era of climate change, the challenge of food security and the uncertainties unleashed by Brexit?

Accounting for the neglect of the social

In a speech to the Oxford Farming Conference in January 2018, the UK government's Secretary of State for the Environment, Food and Rural Affairs, Michael Gove, lavished praise on small farmers:

> There are any number of smaller farm and rural businesses which help keep communities coherent and ensure the culture in agriculture is kept healthy. Whether it's upland farmers in Wales or Cumbria, crofters in Scotland or small livestock farmers in Northern Ireland, we need to ensure support is there for those who keep rural life vital.
>
> (Gove, 2018)

This was one of the very rare occasions when a UK government minister appears to have considered a broader social and cultural agenda related to agriculture. But what might the 'social' mean in the context of agricultural sustainability? There have been some earlier policy attempts to grasp this, mainly focusing on the concerns of food consumers, as though somehow 'social' when applied to agriculture has little relevance to farming people. For example, the Policy Commission on the Future of Farming and Food produced a report in the aftermath of the foot and mouth epidemic, following the orthodoxy of sustainability thinking, and grouped their recommendations under economic, environmental and social headings (Curry Report, 2002).

The section of social recommendations under the label 'People' is the shortest of the three and is concerned entirely with non-agricultural people, specifically focusing on consumers and their concerns (Winter, 2004). The following topics were covered: food labelling, public access to farmland, farm animal welfare, use of antibiotics in animal production, healthy eating and nutrition, public procurement policies and food in schools, and food deserts. These are all highly important issues. They are all of central relevance to agriculture's economic sustainability. But it seems curious indeed to focus discussion of the sustainability of a particular sector almost exclusively on issues of consumption, important though these may be, while neglecting the producers whose practices so determine sustainability outcomes in the field if not on the fork.

One of the arguments in this book, therefore, is that understanding the social in the context of farmers is vital to developing a way forward on food security and agricultural sustainability. In that context, we need to ask who are today's farmers and land owners, as we do in Chapter 3. How do they see their role and how connected are they to the wider drivers of policy, consumer demand, science and technology (Chapters, 4, 5 and 6)? In answering these questions we will range into areas not normally associated with studies of agriculture. For instance, the new emphasis on well-being and happiness leads us to look at both the role of 'happy' farmers in responding to the new challenges and possibilities but also to the wider social 'offer' that agriculture makes, as society as a whole seeks to readjust to the realities of a less carbonised society (Chapter 7). We argue that as land, and how it is used, is so central to tackling climate change, we cannot afford to neglect the social dimension of agricultural sustainability. A sustainable planet needs exemplary land use and that can only be provided by 'happy' land managers.

A broad definition of social objectives within sustainable farming policies and discourses remains hard to find and the suggestions given by Bowler (2002) – that the social dimension of sustainable development in agriculture should cover issues such as the optimum level of farm population, quality of farm life, and the distribution of material benefits – remain largely unexplored (although see Alston, 2004, for some insights).

So why the neglect, why even the lack of a suitable language to explore the social and cultural dimensions of English farming? The main explanation for this lays in England's peculiar agrarian history and the relative economic (un)importance of the agricultural industry in one of the earliest industrialised countries. In contrast to most of Europe, England in the nineteenth century did not have a large agricultural peasantry whose concerns featured strongly in domestic politics. None of the political parties in Britain are rooted in an agrarian peasant politics. Appealing to peasant and ruralist values and concerns has not been a feature of mainstream British politics. Moreover, the extent that agrarian concerns have influenced political parties at all has been long diluted as a result of the tri-partite system of agrarian capitalism built around landlord, tenant and worker. Inevitably the different, indeed conflicting, interests of these three groups prevented the emergence of a strong political or public consensus on what is of social value in the agrarian world. Indeed, of the three groupings it was the farmers who were able to make the biggest political strides in the twentieth century, through the operation of the National Farmers' Union. They did so by abandoning claims to anything socially or economically distinctive about agriculture, instead allying themselves to the political consensus that emerged in the middle decades of the century around the virtues of a managed economy (Cox et al., 1991; Winter, 1996). From the 1870s until 1939 (with the exception of the years of the 1914–18 war) domestic agriculture suffered as result of free trade ideology (Perry, 1973; Turner, 1992; Whetham, 1978). And at the same time, changes in taxation, particularly death duties, and agricultural tenancy law resulted in a decline in the proportion of land let by landlords to tenants and a corresponding rise in owner-occupied farming (Cannadine, 1990; Sturmey, 1955).

The political case for agriculture in Britain during much of the twentieth century revolved around agriculture's place in a modern mixed economy, particularly its contribution to import saving and the gross national product. There were even periods when one of agriculture's economic contributions was seen to be the supply of surplus labour to the urban-industrial complex. In such a context it was hardly surprising that powerful social arguments about agriculture's value to society did not develop in marked contrast to countries such as France with its long history of peasant farming (Mendras, 1976). However, there are shadows of a ruralism even in England. As Winter (1996) shows, out of war-time political turbulence and national emergency, a rejuvenated agriculture was born in the 1940s and 1950s and key to this was the idea of *family* or *working* farming which emerged as result of a particular conjuncture of cultural, political and socio-economic forces. As examined by Winter (2018), culturally, the notion of family farming was promulgated in the period roughly from the 1930s to the 1950s in literary representations in a manner quite unparalleled before or since. In the same period, a politics of agricultural support was forged which saw family farming as a means of circumventing the traditional tri-partite class-based politics of landed capitalist agriculture. Economically, family farming emerged as a result of the relative fixity of the landholding structure along with the declining size of the hired labour force due to mechanisation. In short, family farming emerged almost simultaneously as a cultural ideal, a political project and a socio-economic reality. But this conjuncture did not in fact lead to a (new) settled notion of family farming with deep societal resonance. And, when the force of the post-war particular economic arguments began to fade, they were rapidly replaced by new agri-environment arguments and policies in which the social and cultural role of farmers scarcely figured at all.

Under the new regime emerging in the mid-1980s, and more strongly after the Common Agricultural Policy (CAP) reforms of 1992, farmers were asked to slow down or even reverse the relentless drive to efficient production. For many, the new emphasis on environmentally friendly farming stood in stark contrast to what had gone before. Certainly, farmers were challenged to behave differently and respond to a quite different set of policy signals. But in one way, there was a degree of continuity between these two important eras of agricultural history. In both, the agricultural industry was outward looking and responding to the needs of an urban-industrial society. The needs had changed but the logic remained the same. Farmers were now responding to society's demands for particular habitats, landscapes and recreational provision, as expressed through the various agri-environmental schemes (see Chapter 8). Of course, farmers benefit from these schemes much as they do from conventional agricultural support policies. And, of course, farmer lobbying groups will do their best to ensure policies are constructed which will allow maximum benefits to farmers. But the arguments they are forced to deploy to justify the policies remain those of the wider public good. The justification has shifted from agriculture's role in the macro-economy to the provision of public goods. However, farming lobbyists remain wary of persuading the public or policymakers that there are any arguments of social justice or social value to be made in the case for agriculture.

By contrast, at the heart of the EU's Common Agricultural Policy, and one of the reasons why the UK has been so uncomfortable with the CAP, has been the importance attached to family farming. As Gray (2000) has demonstrated, the notion of family farming provides a crucial unifying symbol that could be bought into by countries with differing notions of family farming and agricultural structures:

> Family farming sustains not just rural society, but society as a whole characterized by the ideals of stability, justice and equality.
>
> (Gray, 2000, p.35)

Although it is possible to find echoes of such ideas in some popular writings about agriculture, particularly in the middle decades of the twentieth century, essentially family farming in the UK has never played this role in the dominant policy discourse. But as agri-environmentalism took hold there did emerge a body of thinking in which social and cultural arguments were more strongly deployed in defence of the countryside. The Countryside Alliance, in its efforts to broaden its arguments and its appeal away from only the defence of hunting with dogs, has sought to develop a new politics which strongly assert the social and cultural significance of country life and traditional country people and mores. No doubt, many would dismiss the Countryside Alliance as inescapably a single issue lobby (concerned with hunting) (Anderson, 2006; Marsh et al., 2009; Wallwork and Dixon, 2004) and ill-attuned to contemporary sensitivities. Nonetheless, its policies and advocacy around such issues as affordable rural housing and rural economic regeneration belie such a characterisation. However, where it fails to strike a chord with contemporary orthodoxies is its relentless championing of the countryside as both coherent, and socially and culturally distinct. For Robin Page, writing in *The Field* (a magazine devoted to the country pursuits of hunting, shooting and fishing) in 1995,

> country people are being swamped, and our land and our communities are being invaded and taken away from us ... country people are being taken over by an increasing wave of urban incomers with urban views, values and culture. As a result I am almost an alien in my own country, lost, persecuted and endangered.
>
> (Page, 1995)

And an editorial, also in *The Field*, in 1998 asks who the Countryside Alliance should represent and makes the following distinction:

> Should the Alliance encompass all country people? Yes, but it depends on the definition of 'country person'. Farmers, foresters, village shop owners, ornithologists and the small boy fishing for rudd in the middle of Newcastle must be in the Alliance. They invest in the countryside and hold a personal stake in the Alliance's aims. Mountain bikers, canoeists and ramblers? They make demands of the land but what do they contribute to its upkeep?
>
> (Young, 1998)

Some might argue that the average rambler contributes a great deal more to the rural economy than the junior angler in Newcastle! Be that as it may, these arguments certainly place the Alliance at odds with a dominant policy discourse which emphasises inter-dependence and the integration of rural and urban. The adoption of social and cultural arguments by the Countryside Alliance has probably done little to assist those who wish to reclaim 'the social' for a progressive programme of sustainable agriculture. But it does at least demonstrate the vacuum in the social discourse associated with agriculture. If we are to focus on and promote the role of agriculture in the transition to sustainability and food security in the era of climate change and environmental degradation, then social arguments matter. Therefore, this book seeks to examine the shifting empirical realities confronted by the farming industry. We seek to offer arguments and interpretation which contribute to building a farming world that is sufficiently self-confident and equipped in terms of knowledge to tackle complex problems.

Sources

The sources used in this book are not confined to our own empirical research. However, we have made particular use of three sources of data from our own research and these warrant a brief introduction here.

SW Farm Survey

The Centre for Rural Policy Research (CRPR) at the University of Exeter undertook three large-scale postal surveys of farmers in the south-west of England (covering the counties of Gloucestershire, Wiltshire, Dorset, Somerset, Devon and Cornwall) in 2006, 2010 and 2016. All three surveys sought basic information about the farmer (age, status in the business) and the farm (size, type, income, type of diversified enterprises, if any), as well as exploring farmers' contact with the community, quality of life, and past and future changes to the farm business (including succession plans). The 2016 survey also included questions on labour, the division of work on the farm, and two questions about Brexit ahead of the EU referendum in June 2016. A total of 1,852 farmers responded to the survey in 2006, 1,543 in 2010, and 1,486 in 2016.

Quantitative data from the surveys were analysed using a statistical computer software package (SPSS). As part of this analysis, we made comparisons between characteristics of sub-groups of respondents using cross-tabulations. In these cases we conducted a statistical hypothesis test for independence between each pair of categorical variables. We calculated a χ^2 (Chi-square) statistic, which measures the *dependence* between two variables, and from this generated a p-value, which is the chance of obtaining such a level of dependence if the two variables were truly independent. A 'significant' association between variables is taken to be one where there is less than a 5 per cent probability of the difference arising by chance ($p < 0.05$). Chi-square values are reported where applicable throughout this book.

SIP Baseline Survey

The Sustainable Intensification Research Platform (SIP), which ran from 2014 to 2017, was a multi-partner research programme funded by Defra and the Welsh government to explore the opportunities and risks for sustainable intensification (SI) in England and Wales (see www.siplatform.org.uk for further information). The CRPR led a major part of the programme (SIP 2). The programme comprised a number of interlinked research projects across seven case study areas covering a range of landscapes and farming systems. The case study areas, based on river catchments (see Figure 1.1), were; the Avon (Hampshire), Conwy (North Wales), Eden (Cumbria), Nafferton (Northumberland), Taw (Devon), Upper Welland (East Midlands), and Wensum and Yare (Norfolk). We draw on a range of findings emerging from SIP research strands throughout this book. In particular, data from the baseline survey conducted as part of SIP 2 ('Opportunities and risks for farming and the environment at landscape scales') is used to inform our discussions around farmers and sustainable intensification (Chapter 5); stress, well-being and community (dis)connections (Chapter 7); and farmer attitudes towards the environment (Chapter 8). The baseline survey was conducted in 2015 and

Figure 1.1 SIP case study areas

consisted of 244 face-to-face interviews with farmers and land managers in the seven case study areas. Using both quantitative and qualitative questions, the survey collected a range of information on current farm characteristics, agri-environmental practices, collaborative activities and quality of life, as well as exploring land manager opinions on a range of topics relating to the sustainable intensification of agriculture (for the full report see Morris et al., 2016). As with the SW Farm Surveys, our analysis of the SIP Baseline Survey data included performing a number of cross-tabulations and testing for significance using the Chi-square statistical test.

Family farming research for the Prince's Countryside Fund

We also draw on work carried out for a report to the Prince's Countryside Fund (PCF) on the future of small family farming in the UK (Winter and Lobley, 2016). This research included specially commissioned analysis of financial performance data for England and Wales from the Farm Business Survey (FBS) 2014/15 (see Wilson, 2016), as well as an extensive literature review; interviews with seven agricultural sector experts; a farmer workshop; and a call for evidence (which received 21 responses).

References

Alston, M. (2004). Who is down on the farm? Social aspects of Australian agriculture in the 21st century. *Agriculture and Human Values*, 21(1), 37–46.

Anderson, A. (2006). Spinning the rural agenda: The Countryside Alliance, fox hunting and social policy. *Social Policy & Administration*, 40(6), 722–738.

Bowler, I. (2002). Developing sustainable agriculture. *Geography*, 87(3), 205–212.

Cannadine, D. (1990). *The Decline and Fall of the British Aristocracy.* New York and London: Yale University Press.

Cm 9577 (2018). Health and Harmony: The future for food, farming and the environment in a Green Brexit. *Presented to Parliament by the Secretary of State for Environment, Food and Rural Affairs.*

Cox, G., Lowe, P. & Winter, M. (1991). The origins and early development of the National Farmers' Union. *The Agricultural History Review*, 39(1), 30–47.

Curry Report (2002). Farming and Food: A sustainable future. Report of the Policy Commission on the Future of Farming and Food. Cabinet Office, London.

Gove, M. (2018). Farming for the next generation. *Oxford Farming Conference 2018.* Oxford, 04/01/18 www.gov.uk/government/speeches/farming-for-the-next-generation [accessed 15/08/18].

Gray, J. N. (2000). *At Home in the Hills: Sense of place in the Scottish Borders.* Oxford: Berghahn.

Helm, D. (2016). British Agricultural Policy after Brexit, Natural Capital Network – Paper 5. Natural Capital Network, London.

Lang, T., Millstone, E. & Marsden, T. (2017). A Food Brexit: Time to get real. A Brexit briefing. Science Policy Research Unit, University of Sussex.

Marsh, D., Toke, D., Belfrage, C., Tepe, D. & McGough, S. (2009). Policy networks and the distinction between insider and outsider groups: The case of the Countryside Alliance. *Public Administration*, 87(3), 621–638.

Mendras, H. (1976). *Les Sociétés Paysannes. Eléments pour une Théorie de la Paysannerie.* Paris: Armand Colin.

Monbiot, G. (2017). The Lake District's world heritage site is a betrayal of the living world. *The Guardian*, 11/07/17.

Morris, C., Jarratt, S., Lobley, M. & Wheeler, R. (2016). *Baseline Farm Survey – Final Report.* Report for Defra project LM0302 Sustainable Intensification Research Platform Project 2: Opportunities and Risks for Farming and the Environment at Landscape Scales.

NFU (2016). Farming's Offer to Britain: How Farming Can Deliver for the Country Post-Brexit. NFU, Stoneleigh.

Page, R. (1995). A persecuted minority. *The Field*, December, p.9.
Perry, P. J. (ed.) (1973). *British Agriculture 1875–1914*, London: Methuen.
Rebanks, J. (2015). *The Shepherd's Life: A tale of the Lake District*. London: Penguin.
Sturmey, S. G. (1955). Owner-farming in England and Wales, 1900 to 1950. *The Manchester School*, 23(3), 245–268.
Turner, M. (1992). Output and prices in UK agriculture, 1867–1914, and the Great Agricultural Depression reconsidered. *The Agricultural History Review*, 40(1), 38–51.
Wallwork, J. & Dixon, J. A. (2004). Foxes, green fields and Britishness: On the rhetorical construction of place and national identity. *British Journal of Social Psychology*, 43(1), 21–39.
Whetham, E. (1978). *The Agrarian History of England and Wales, Vol. VIII 1914–1939.* Cambridge: Cambridge University Press.
Wilson, P. (2016). The Viability of the UK Small Farm: Analysis of Farm Business Survey 2014–15 data for England and Wales. Specially commissioned report for Prince's Countryside Fund small farm research. Centre for Rural Policy Research, University of Exeter.
Winter, M. (1996). *Rural Politics: Policies for agriculture, forestry and the environment.* London: Routledge.
Winter, M. (2004). Who will mow the grass? Bringing farmers into the sustainability framework. *Journal of the Royal Agricultural Society*, 165, 113–123.
Winter, M. (2018). (in press) Farming Tales: narratives of farming and food security in mid-twentieth century Britain. *In:* Mukherjee, A. (ed.) *Food Security and the Environment in India and Britain.* London: Routledge.
Winter, M. & Lobley, M. (2016). *Is there a Future for the Small Family Farm in the UK?* Report to the Prince's Countryside Fund. London: Prince's Countryside Fund.
Young, J. (1998). The real power behind the Alliance. *The Field*, September, p.9.

2 Sustainable agriculture in a world of food security

Introduction

In this chapter, we review some of the background to the changing world of agriculture in a global context. Although most of this book is about the specific case of farmers in the south-west of England, these farmers operate against the backcloth of global forces and challenges and here we summarise the key issues. In particular, we suggest that the twin requirements of farming *sustainably* and *productively* constitute an unprecedented challenge to the agricultural industry; a 'wicked' problem that cannot solely be solved by the application of the traditional approaches of either agricultural science or agricultural economics. In 2009, this challenge was presented graphically as a 'perfect storm' by the then UK government's Chief Scientific Advisor Sir John Beddington, in which he set out a worrying future based around interconnections between the three resources of food, water and energy (now often called the nexus) in the context of climate change (Beddington, 2009). This was followed in 2011 with the highly influential World Economic Forum's Global Risks report echoing Beddington's warnings: 'any strategy that focuses on one part of the water-food-energy nexus without considering its interconnections risks serious unintended consequences', identified in this case as risks to both economic growth and global security (World Economic Forum, 2011, p.7). The timing of these initiatives is important, following so closely the food price increases of 2008 which we discuss in more detail later in this chapter.

For farmers, the emergence of food security as a policy driver heralded a sudden broadening of public concern beyond the environmental concerns that had dominated for the previous two decades. This was potentially encouraging news for food producers, promising the possibility of a significant departure in policy approaches to the industry. This was not lost on farmers, although our research has shown that many in the south-west of England have ambivalent interpretations of the food security agenda due to scepticism about the failings of a highly industrialised and globalised agro-food system and its varied dependencies, inefficiencies and risks (Fish et al., 2013). The requirement to integrate and harmonise economic, social and environmental objectives promises the possibility of a broader policy approach than the one that has been visited upon the industry in the past. There are pivotal moments in the twists and turns that have characterised the development of agriculture, and the policy framework in which it operates in the UK, in the last century. Two world wars, especially the second, ushered in a reappraisal of the importance of agriculture to the nation; the legislative programme of the 1945–50 Labour government reinforced this reappraisal but at the same time broadened the rural canvass beyond farming to also bring land use planning and environmental protection to centre stage (Flynn, 1986; Sheail, 1997). To hugely simplify, the first three and a half post-war decades were characterised by a UK agricultural policy dominated by economic concerns; thereafter environmental issues came more to the fore (Winter, 1996). During both phases, social concerns were low on the policy agenda. The economic success of the industry from the 1940s through to the 1970s was achieved at a social cost, with labour shedding and a decline in the number of farms. Tensions implicit in this post-war settlement eventually surfaced giving rise to a further policy turn in the early 1980s, with measures designed to restrain agriculture and protect a diminishing environmental resource. We entered into what has been termed a post-productivist policy era, though this is a contested concept, critiqued for its inconsistencies and overly simplistic dualism (Evans et al., 2002; Wilson, 2001); a 20-year period during which the policy emphasis could best be described as agri-environmental and domestic self-sufficiency in food production declined (see Figure 2.1). This was abruptly challenged by global events in 2008.

The growth of environmental regulation in the 1980s and 1990s, which many would argue was entirely justified on the grounds of demonstrable habitat loss and resource depletion, was implemented with insufficient regard for the social costs as farmers struggled to come to terms with a volte-face in public expectations. Whereas the literature on environmental impacts of agriculture is voluminous (overviews are available in Foster et al., 2006; Chamberlain et al., 2000; Skinner et al., 1997), coverage of the social impact of agricultural change on rural and farming communities is slim. This neglect of the 'social' has become central to the current agricultural policy agenda for the simple reason that increasing food production in a sustainable manner is profoundly challenging. Economic incentives alone are insufficient and, in any case, are hard to imagine in the current neo-liberal context.

Farming and the rise of food security: demand-side context

The food price spike of 2008 (see Figure 2.2) came as a real surprise to many in policy circles in the UK, long used to liberal or neo-liberal assumptions about the efficacy of free trade arrangements in providing food at low prices through global trade. The reasons for this rapid escalation of prices have been, and continue to be, keenly debated. On the supply-side, poor wheat harvests and consequent lower grain stocks, and a rise in oil prices (Baumeister and

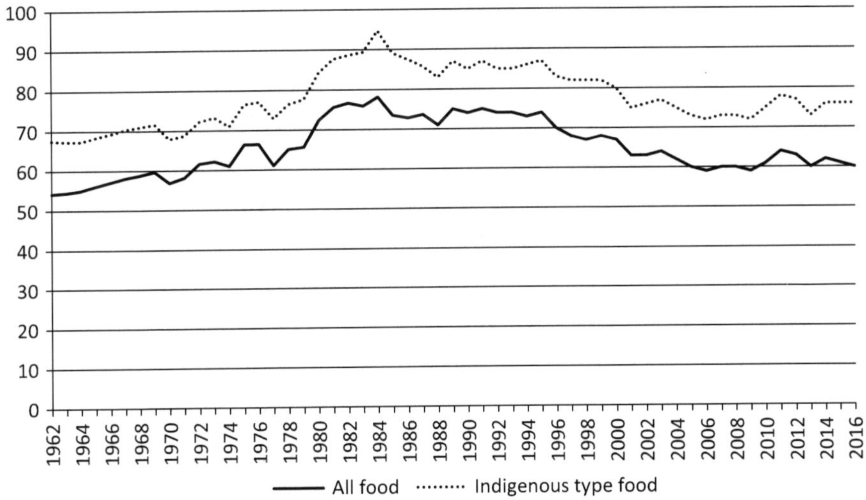

Figure 2.1 UK food production to supply ratio (commonly known as the self-sufficiency ratio), 1962–2016

Note: Figures for 1962–1988 are a merged series where the trend estimate has been retained but data adjusted to bring it in line with a new methodology which was introduced in 1998 and applied to data from 1988. The new methodology applied revaluation factors to imports as well as exports.

Source: Defra, 2014, 2017b

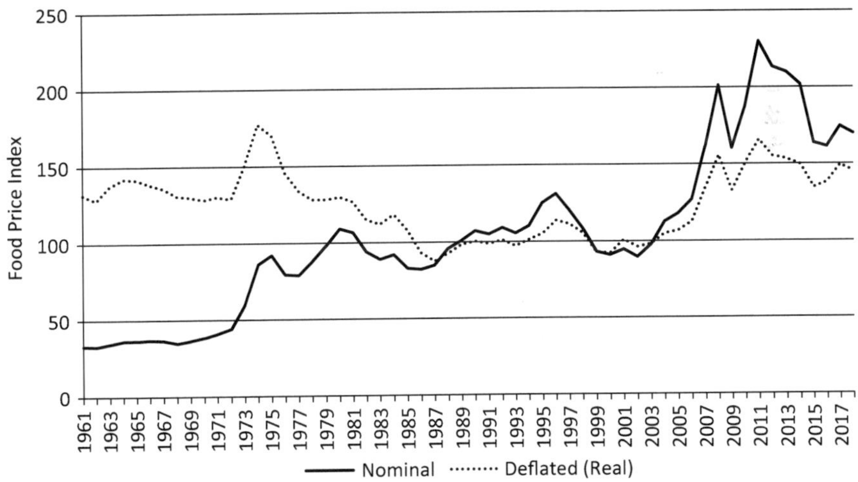

Figure 2.2 FAO Food Price Index in nominal and real terms, 1961–2018

Source: FAO, 2018

Kilian, 2014), are key elements in the debate. On the demand-side, contributory causes included inflation arising from the rapid growth of the world economy in the early 2000s and depreciation of the US dollar (Wiggins et al., 2010). The export bans and restrictions, restocking in tight markets and reduced import tariffs that were policy responses exacerbated the situation (ibid). Rather more contentious than these were the possible effects of a diversion of grains to distillation for biofuels (Rosegrant and Msangi, 2014), rising demand for food in China, India and other emerging economies and, perhaps the most contentious of all, speculation on futures markets. Tadesse et al. (2014) have investigated the main drivers of food price spikes for wheat, maize and soybeans, and show how exogenous shocks as well as the linkages between food, energy and financial markets play a significant role in explaining food price volatility and spikes.

Variations in the Food Price Index highlight the fluctuating nature of the market and reflect the impact that changes in global conditions (climatic, economic, political) exert over international prices. Furthermore, this occurs on not just an annual but a monthly basis. For instance, a rise in the Food Price Index in summer 2017 was driven largely by increases in dairy and cereal prices between April and July which, in the case of dairy products (particularly butter), was due to limited export availabilities in all major producing countries. Cereal prices rose in this period despite downward pressure on maize prices (due to record harvests in South America) as a result of large increases in wheat prices following poor crop conditions in the United States. The overall increase in the 2017 index also conceals a large drop in sugar prices between February and July due to high export availability (particularly from Brazil) and weak import demand due to the imposition of high import tariffs by China in May 2017 (Food and Agriculture Organization, 2017). Such price volatility is inevitable in a globalised market but results in high levels of uncertainty for farmers, many of whom (particularly those with smaller units) consequently struggle to plan accordingly and make sufficient return on their products. As we discuss in Chapter 7, this not only affects the viability of the farm business but, as a significant source of stress, has a negative impact on farmer well-being.

Global food demand in 2050 is projected to increase by at least 60 per cent above 2006 levels (Food and Agriculture Organization, 2016) as a result of a rising world population, income growth and urbanisation. As depicted in Figure 2.3, we already live in a world of 7.5 billion people (60 per cent of whom live in Asia; see Figure 2.4) and this is projected to rise to 9.8 billion by 2050. According to Food and Agriculture Organization (FAO) statistics, there are enough calories being produced in the world today to feed the entire population: in 2011, 2,870 kcal per capita per day were available for human consumption at the global scale (Food and Agriculture Organization, 2015). One of the problems is that those calories are not being distributed evenly, with 868 million people going hungry while 1.4 billion are overweight or obese (Food and Agriculture Organization, 2013). Quite apart from the obvious

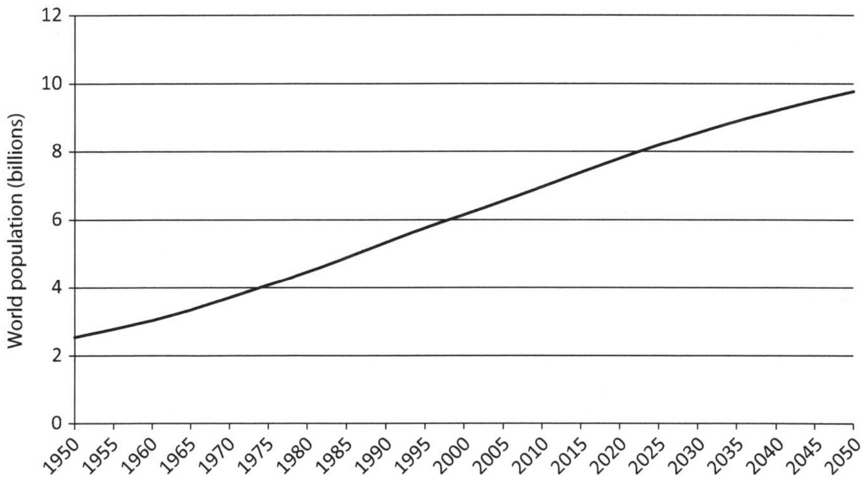

Figure 2.3 World population, 1950–2050

Source: United Nations, 2017

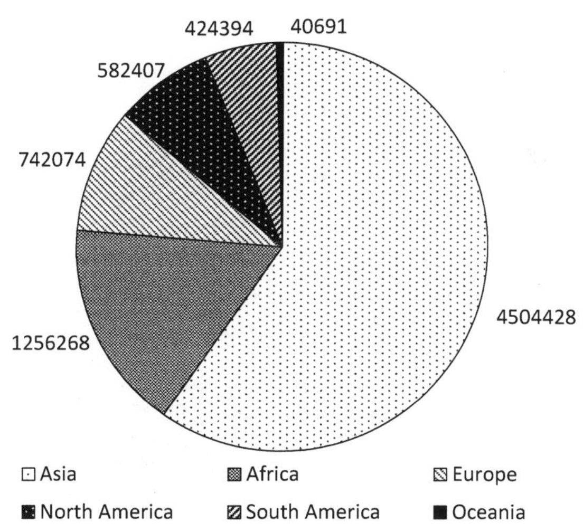

Figure 2.4 World population by continent, mid-2017

Source: United Nations, 2017

issues of food justice and equity implied by these figures, overconsumption has significant negative implications for human health and the economy.

Projected rising demand caused by a growing population is further exacerbated by growing purchasing power. Higher incomes in countries such as China, Brazil and India are driving an increased demand for food, particularly meat and dairy products. These dietary preferences for meat and dairy compound food security issues because livestock production requires large amounts of land and feed-crops and, in theory, more people could be supported from the same amount of land if they were vegetarians. As Godfray et al. (2010) point out, however, the situation is more complex than that argument suggests, as there are wide variations in production efficiency between different types of meat and much of the grassland that is used to feed livestock could not be converted to arable land without significant environmental impact. Nevertheless, a potential doubling of global meat demand by 2050 (The Royal Society, 2009) poses a substantial challenge for sustainable food production. This 'global nutrition transition' (Popkin et al., 2012) (see Figure 2.5) is characterised by shifts towards a 'Western diet', which includes a high intake of refined carbohydrates, added sugars, fats and animal-source food (partly in the form of caloric beverages, processed and pre-cooked food),

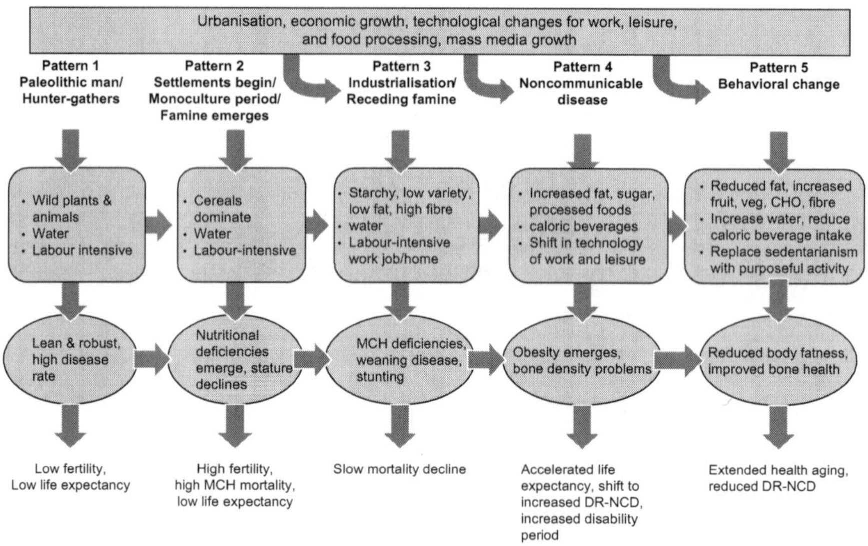

Figure 2.5 Stages of the nutrition transition

Notes: DR-NCD = Diet-related non-communicable disease. NCD is a medical condition or disease that is not caused by infectious agents and therefore is non-infectious or non-transmissible.

Source: Popkin, 2006. *Global nutrition dynamics: the world is shifting rapidly toward a diet linked with noncommunicable diseases* [1–3], The American Journal of Clinical Nutrition, Vol.84 (2), p.291. By permission of Oxford University Press

alongside a reduction in fruit and vegetable intake. These changes clearly have major implications for agriculture around the world although, as we explore in more detail in Chapter 4, the situation in the UK is more nuanced. With regard to ruminant livestock and dairy farming (the dominant enterprises in the south-west of England – see Chapter 3), domestic UK consumption is static or declining, depending on product, as diets change. Whereas globally the reverse is the case; hence the interest in the export market shown by organisations such as the National Farmers' Union (NFU) in the Brexit debate.

Nor should it be assumed that the global nutrition transition is only significant in terms of increasing demand for food products therein fuelling agricultural growth. On the contrary, the human health concerns associated with dietary change and declining physical activity in developing countries (partly due to technology leading to less energy-intensive work activity) mean that diet is increasingly under the spotlight with all that that might mean to long-term trends in demand for foodstuffs. The basic characteristic of the Nutrition Transition is an increase in the consumption of calories but the change is much deeper and more significant than that and includes:

- Large increases in consumption of caloric beverages and snacking;
- Increased intake of ultra-processed foods;
- Greater consumption of refined carbohydrates;
- Reduced intake of fruits and vegetables and legumes;
- Reduced preparation time and increased use of pre-cooked foods;
- Too many empty calories and not enough vitamins, minerals, antioxidants, amino acids or fibre.

The consequences of the changes for human health are evidenced most graphically in the increase in obesity worldwide, as shown in Figure 2.6, which contributes to an increasing incidence of non-communicable diseases (NCDs) such as cardiovascular diseases (heart attacks and stroke), cancers, chronic respiratory diseases (such as chronic obstructive pulmonary disease and asthma) and diabetes. The role of dietary change in these is now widely accepted (Global Burden of Metabolic Risk Factors for Chronic Diseases Collaboration, 2014) even if precise causal mechanisms are harder to pin down. The FAO (2013) predicts that the costs of obesity to the global economy in the form of healthcare and loss of productivity could be as much as 5 per cent of global GDP. Furthermore, problems of obesity are no longer limited to the developed world and are linked to global trends concerning food habits and exercise levels. Such changes inevitably lead to negative impacts on health (and the economy), with nutritional transitions in countries such as China and India being linked to rises in NCDs (Popkin et al., 2001). In 2015, 40 million (70 per cent) of the 56 million global deaths were due to NCDs, with over three-quarters of these deaths occurring in low- and middle-income countries (World Health Organization, 2015).

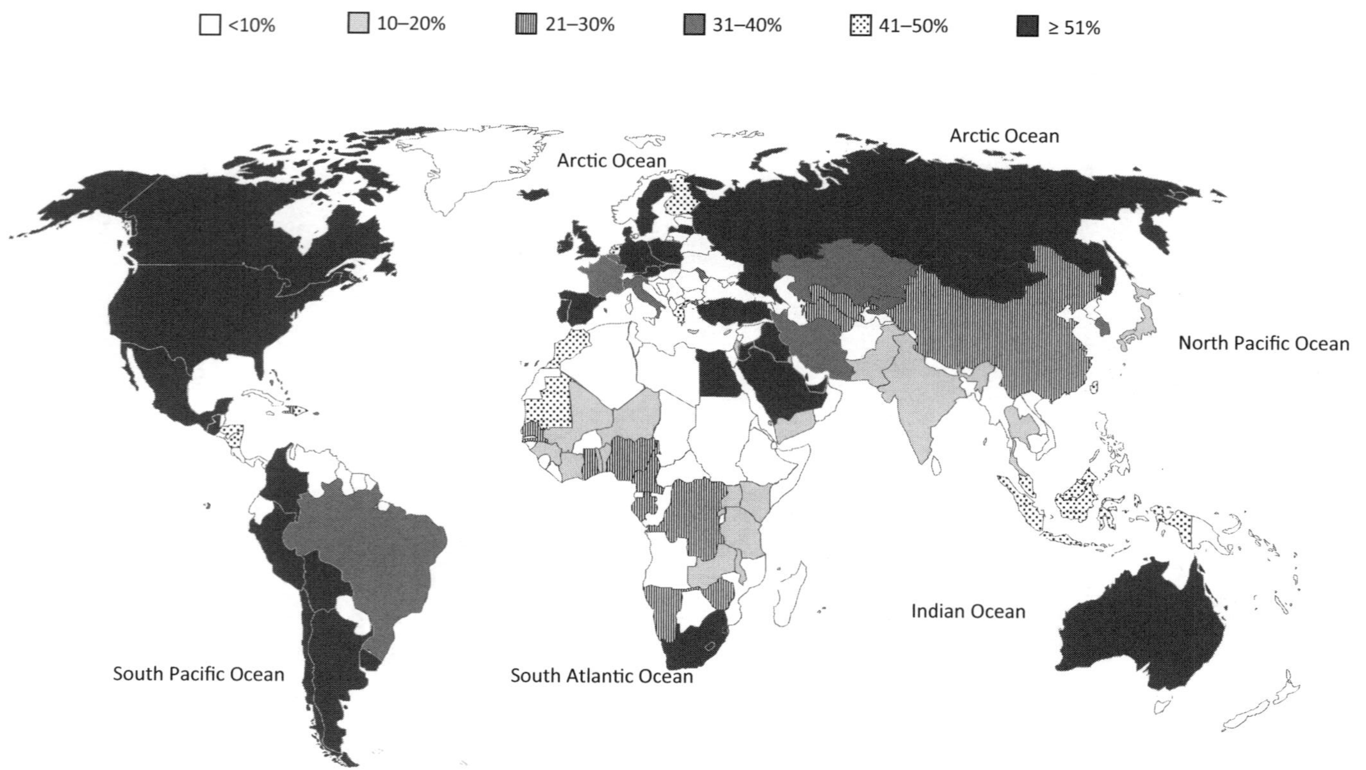

Figure 2.6 Patterns of overweight and obesity globally for nationally representative samples

Source: Popkin, 2012, by permission of Oxford University Press USA. *Food and Addiction: A Comprehensive Handbook* by Brownell and Gold (2012), Fig.10 from Chp. "The Changing Face of Global Diet and Nutrition" by Popkin. By permission of Oxford University Press USA

As Winson (2004; 2013; Winson and Choi, 2017) has argued, diets are ultimately social, economic and political as much as they are personal and behavioural. Diets reflect the material conditions of a particular society, and specific social and economic arrangements, as well as the structures of political domination, regulation and control (Winson, 2013, p.16). The concept of *dietary regime*, developed by Winson as an alternative to the nutrition transition, redirects the focus of analysis in an attempt to capture

> the commonality of dietary experience and to guide an appraisal and understanding of the social forces and socio-economic and technological factors that play a salient role in determining prevailing diet(s) in a society at a given point in time.
>
> (Winson and Choi, 2017, p.564)

Two issues flow from this discussion with regard to the main focus of this book on *agriculture*. First, food security is no longer, if indeed it ever was, solely about the *availability* of food. It is now also about nutritional quality and social justice. As Pinstrup-Andersen (2009) pointed out in the inaugural issue of the journal *Food Security*, the use of the term food security at the national and global level focuses on the supply-side. However,

> The question raised is: is there enough food available, where food is usually interpreted to mean dietary energy? But availability does not assure access, and enough calories do not assure a healthy and nutritional diet. The distribution of the available food is critical. If food security is to be a measure of household or individual welfare, it has to address access.
>
> (Pinstrup-Andersen, 2009, p.5)

For farmers, this means the potential for a heightened attention to the nutritional content of their products extending far beyond the traditional concerns of safety, quality and provenance (explored in greater detail in Chapter 4). We have seen signs of this already in an increased interest in research into nutrition-sensitive agriculture interventions, in various contexts across the world (Berti et al., 2016; Yu and Tian, 2018) and to some extent in the UK (Freitag et al., 2018). In short, as a result of the health consequences of the nutrition transition, farmers may face new market changes presenting both challenges and opportunities (see Winter, 2018, for a longer consideration of this).

Second, farmers may also face policy challenges as inevitably attempts will be made to reduce the burden of escalating health costs on the public purse. In the UK obesity has the second largest health impact after smoking, generating an economic loss of more than US$70 billion (c. £50.4 billion) a year in 2012, or 3 per cent of GDP (Dobbs et al., 2014). Currently, the typical policy approach to the obesity challenge is primarily based on the need for behavioural change among consumers and within food supply chains. Thus the McKinsey Global Institute (Dobbs et al., 2014) suggests 18 areas of intervention, covering *inter alia* physical exercise, food advertising and labelling, education and public health initiatives. Just one directly concerns agricultural subsidies, taxes and prices, and the following suggestions are made:

Relative price increase: regulated	→	Government introduces a tax in order to drive price increases on certain types of food or nutrient
Relative price increase: reduced agricultural subsidy	→	Government reduces subsidies on certain food commodities that drive prices (e.g. processed foods such as corn, sugar and palm oil)
Relative price decrease on fresh produce and staple foods: increased agricultural subsidy	→	Government subsidises fresh food such as fruit and vegetables
Relative price decrease on fresh produce and staple foods: personal subsidies	→	Government provides personal subsidies (e.g. food stamps for low-income individuals for sole use on certain healthy food types)

Similarly, Hawkes et al. (2013), for the World Cancer Research Fund International, argue how food system and nutrition policies will inevitably impact on agricultural policy as attempts are made to harness food policies and actions across sectors to ensure coherence with healthy eating:

> The rationale for taking policy action in the food system and supply chain has three inter-related aspects. The first aspect is that, through their effects on food availability, affordability and acceptability, specific agricultural and food systems policies have repercussions for policies to promote healthy eating. For example, trade policies that make it easier to obtain ingredients used in high calorie foods may conflict with policy actions taken to reduce their consumption. Identifying these repercussions and creating policy coherence – a 'health in all policies' approach – could increase the effectiveness and sustainability of the policy actions. This would require creating the governance structures to develop such actions. The second aspect is that policies that address the food environment have inevitable repercussions upstream for the actors and activities in agriculture and food systems. Governments thus need to anticipate and respond to the reaction of the food supply chain. ... The third aspect is that policy actions can be implemented in food systems with the explicit intention of changing food availability, affordability and acceptability to promote healthy diets. Improving the logistics of supply chains to make culturally acceptable fruits and vegetables available through public procurement is a case in point.
>
> (Hawkes et al., 2013, p.14)

Farming and the rise of food security: supply-side context

Food security is often seen primarily as a demand-side issue with population and diet, as discussed in the previous section, the key concerns. But food security and the farming response to demand is also dependent on supply-side factors. Farmers' ability to respond to demand is dependent on the cost and availability of production inputs, land availability, weather/climate and technological developments. And in this respect farmers face considerable uncertainty. Supply-side challenges include pressures on key resources such as oil, water, nitrate, phosphates and soil; land use pressures; a declining growth in agricultural productivity (discussed in Chapter 5); and climate change. These are all massive topics and each one has generated large literatures. Here we merely touch on some of the most salient points particularly as they relate to our focus on the south-west of England. We do so in the context of declining resource availability inevitably having an economic implication, raising the costs of agricultural inputs for farmers.

Soil nutrients

The four major nutrient inputs needed for plant growth are nitrogen, phosphorus, potassium and sulphur. Potassium and sulphur are not expected to be limited in supply in the foreseeable future nor, crucially, is there a significant energy requirement in their processing. There are, of course, additional micro-nutrients that are also essential, such as iodine and selenium, which are so depleted in UK soils that grazing animals usually need supplements (Bowley et al., 2017) and the role of the soil microbial community is also important if not yet fully understood (Corstanje et al., 2015; Griffiths et al., 2008). The depletion of soil organic matter may also be associated with compaction and erosion in many soils (Gregory et al., 2015). But soil biology is not fundamentally dependent on industrial processes and with good management can be improved over time. By contrast, nitrogen and phosphorus (hereafter N and P) pose a serious challenge for two very different reasons: in the case of N because of the costly and energy demanding process used to manufacture nitrogenous fertiliser; and the in the case of P because rock phosphate, from which manufactured P is derived, is a finite resource (although there are also resource and energy requirements too, as P fertilisers are manufactured by burning sulphur to make sulphuric acid for the dissolution of phosphate rock, and most sulphur used for P fertilisers is supplied by the oil and gas industry – Dawson and Hilton 2011).

Turning to N first, there are, of course, unlimited supplies of N in the atmosphere but this inert form of N is not available for plant growth other than for legumes and a few other plants that have root nodules that can fix N. And some soil microorganisms convert organic forms of nitrogen to plant-available mineral forms when they decompose organic matter and fresh plant residues through mineralisation. Some plant-available N is provided by rain water but 98 per cent of the nitrogen in soil is in organic forms. None of these sources are sufficient for the levels of production usual in what have become *conventional* agricultural production systems. Note the emphasis given to the word *conventional* – some have claimed that natural fixation by legumes could provide enough N to maintain current high levels of food production if conventional systems of agriculture were replaced by organic systems (Badgley et al., 2007). Others disagree (Goulding et al., 2009). Either way, the current food system is highly dependent on the relatively affordable availability of N fixed industrially by the Haber–Bosch process whereby atmospheric nitrogen (N_2) is converted to ammonia (NH_3) by reaction with hydrogen (H_2), using a metal catalyst under high temperatures and pressures (Vaclav, 2004). Although the process is widely regarded as efficient, it does require significant energy inputs to maintain temperature and supplies of methane gas for the hydrogen. This leads Dawson and Hilton (2011) to suggest that, long term, methane might need to be conserved and ring-fenced from general energy provision to be used solely within the Haber–Bosch process since, albeit less conveniently, the energy deficit left by deploying methane for this purpose can be made up by other means. Its importance is undeniable:

> Calculations of the N flows in global agriculture carried out by Smil (1999) estimated that '40% of the world's dietary protein supply in the mid-1990s originated in the Haber–Bosch synthesis of ammonia'. Smil concluded that about 85% of the N in food proteins available for human consumption, both plant and animal, were derived from cultivated crops, with the balance from extensive grazing and fish. Given that half of the N in harvested crops is derived from the N fertiliser applied, by the mid-1990s some 40% of the world protein supply came from N fertilisers. Erisman et al. (2008) revisited the calculations, concluding that the figure is now 48%, a marked upward trend. … Industrial production of reactive N is essential for feeding a large and increasing global population, and the most efficient production method is methane-based (Jenssen and Kongshaug, 2003; Pach, 2007). A long-term strategy for global food security depends critically on N availability, which in turn requires that adequate supplies of methane are reserved for this function.
>
> (Dawson and Hilton, 2011, pp.S16–S18)

Turning now to P, estimates suggest that world rock phosphate production will peak in around 2030 (see Figure 2.7) and that current global reserves may be exhausted in the next 100 years (Cordell et al., 2009; Heffer et al., 2006; United States Geological Survey, 2010), 125 years (Smit et al., 2009) or, rather more optimistically, 300–400 years (International Fertilizer Development Center, 2010). The biggest reserves are to be found in Morocco, more than two-thirds of global reserves by some calculations, and there is virtually none in western Europe:

> The UK, like most of Europe, has no phosphate rock reserves of its own and therefore relies on imported mineral phosphate fertilisers to support UK agriculture. Phosphorus is also imported within food and animal feed. Therefore, UK food security is currently highly dependent on a

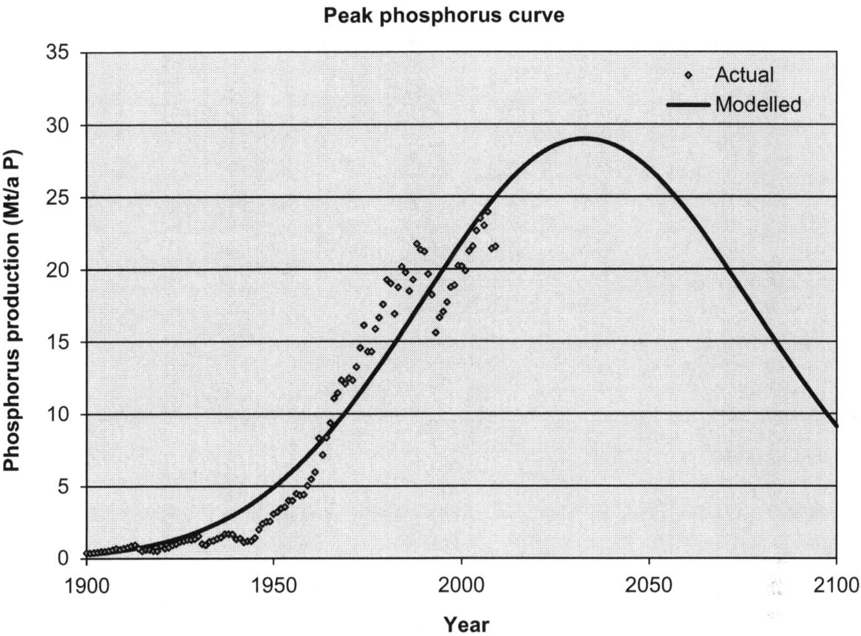

Figure 2.7 Indicative peak phosphorus curve

Source: Reprinted from Cordell et al., 2009, with permission from Elsevier

> secure, affordable supply of mineral fertilisers derived from phosphate rock. In order to increase UK food security, the UK could move towards a more resource efficient, closed-loop system which reduces overall consumption, prevents losses and recycles nutrients with greater efficiency.
>
> (Cooper and Carliell-Marquet, 2013, p.83)

Soils

Between 2002 and 2011, the Environment Agency carried out research into the degradation of soil structure in the South West at over 2,500 sites in 21 catchments (Figure 2.8):

> Overall, 38% of sites showed high or severe degradation with signs of erosion and runoff, 50% displayed moderate damage, and only 10% had low levels of damage. Cultivated sites posed a large problem, with 55% having high or severe damage, while less than 10% of permanent grass sites had high or severe damage. Fields with maize or potatoes, or other late-harvest crops, were the most damaged, with 75% of those sites showing degradation. In fact, one in five of those sites had serious rill and

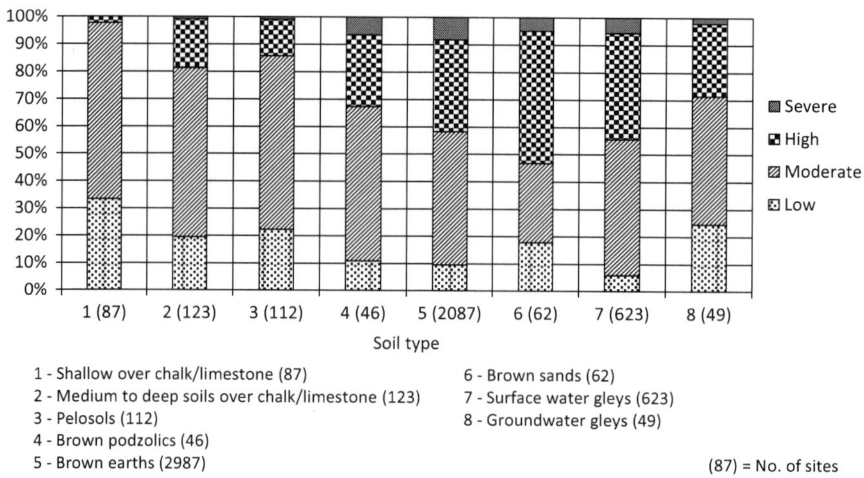

Figure 2.8 Soil degradation within broad soil landscapes in the South West, 2002–2011

Source: Palmer & Smith (pers. comm.)

> gully erosion. They found that winter cereal crops, like wheat and barley, also caused problems, with 60% of sites from those fields displaying high or severe degradation.
>
> (Palmer and Smith, 2013, p.567)

Maintaining soil resilience is highly challenging and the science complex (Corstanje et al., 2015). Policy is weakly developed and there are growing calls for more policy attention to this issue, for example a national-scale soil conservation and soil condition framework championed by Humphries and Brazier (2018).

Competition for land

Food production also faces challenges in the form of competing demands for land. Given the high population density in the UK, especially in England, land competition and as a result land prices (we discuss the price of land in Chapter 3), are high. To some extent, the production increases needed to meet demand may be achieved through technological and efficiency improvements within agriculture, but it is likely that additional land will also be required. However, levels of urbanisation, biofuel production and industrial forestry are also set to significantly expand over the coming decades, making land an increasingly scarce resource (Lambin and Meyfroidt, 2011). The production of biofuels and the 'food vs fuel' debate is perhaps the most well known of these issues following rapid increases in recent years, particularly in the United States

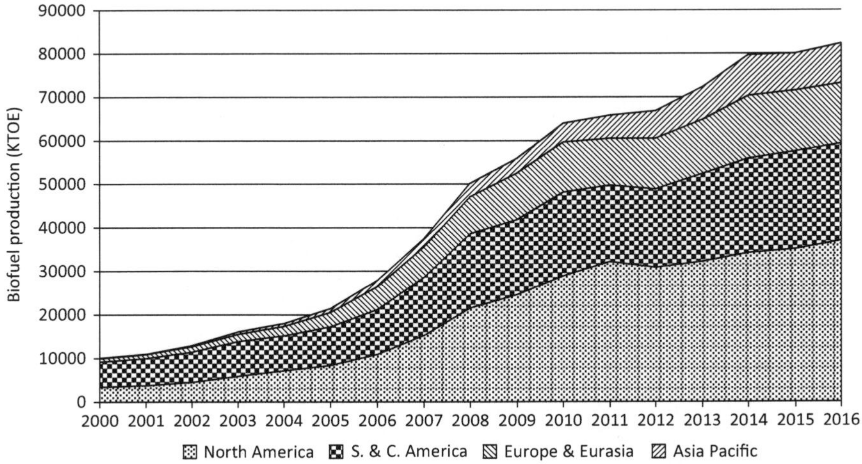

Figure 2.9 Global biofuel production 2000–2016 in kilo tonnes of oil equivalent (KTOE)

Source: BP, 2017

and Brazil (see Figure 2.9) (Baines, 2015). In the UK, while biofuels have not expanded in the way they have elsewhere in the world, there has been an increase in maize production destined for use in anaerobic digestion (Farnworth and Melchett, 2015). Environmental impacts associated with land conversion for arable crops (for both food and non-food uses) are also of high concern.

Montague-Fuller (2014), for the Cambridge Natural Capital Leaders Programme, undertook an analysis of demand and supply options for UK land to 2030, assuming a population growing from 62.6 million in 2012 to 71.4 million by 2030 and no significant change in the balance of UK food trade with global markets (the work was undertaken before Brexit!). The following additional demands were assessed:

- Increased residential land needed for a larger UK population by 2030;
- Land needed to meet bioenergy targets under Department for Energy and Climate Change 2050 scenarios;
- Improved UK food security through replacing some key imports where viable, such as apples, pig meat and some high protein animal feed; and increasing exports where UK has a competitive advantage, such as root vegetables, cereals and oilseeds;
- Improved wildlife and habitat protection through increased areas managed for nature;
- Greater woodland cover to deliver a range of benefits including carbon storage, wood fuel, flood management, increased wildlife and new amenity areas;

- Land dedicated to improved water management infrastructure, such as increased wetlands and new reservoirs.

(Montague-Fuller, 2014, p.10)

On the supply-side, the following possibilities that might reduce the amount of land needed for agricultural production were considered:

- Sustainable intensification opportunities, including precision farming and improved seed varieties. This could deliver higher yields that enable arable land to be released for other uses;
- Likewise, livestock yield increases, including improved husbandry, collection and use of performance data, and better feed conversion ratios. These will allow the same meat and dairy production to be delivered from less land. Current government and industry cooperation on agritech research is helping to advance these types of opportunities;
- Reductions in food waste, which enable less land to be used to feed the same population. Approximately 19 per cent of household food and drink is currently wasted in households;
- Change in UK eating habits, with a move from meat to non-meat-based diets in line with the UK trend over the last 20–30 years. With livestock requiring up to 10 times the area to produce the equivalent amount of protein for human consumption as arable crops, a reduction in meat consumption could release a significant amount of land for other uses (except where livestock are being reared for the global market).

(Montague-Fuller, 2014, pp.10–11)

The conclusions are striking and suggest that land use pressures will only grow:

> Taking the high demand case and low supply case as an extreme, a land shortfall of approx. 6 million hectares could occur, equivalent to a third of current UK agricultural land. Prima facie, this could lead to some very difficult choices having to be made on UK land use. Nevertheless, it is acknowledged that there is a high degree of uncertainty over the extent of this demand-supply imbalance, particularly given the lack of data and associated targets for the majority of the demand and supply categories shown.
>
> (Montague-Fuller, 2014, p.12)

The high degree of uncertainty of course bedevils this sort of approach, helpful though it may be in highlighting issues.

Climate change

Climate change has a number of implications for future world food security. Impacts on food production include the immediate effects of extreme weather (flooding, drought, storms etc.) on crops and infrastructure for food producers, particularly in vulnerable regions, as well as longer-term risks associated with sea-level rise and reduced availability of water. Consequences for agriculture may be acute, particularly for producers in the developing world: the FAO predicts that 'with climate change, the population living in poverty could increase by between 35 and 122 million by 2030 relative to a future without climate change, largely due to negative impact on incomes in the agricultural sector' (Food and Agriculture Organization, 2016, pp.9, 13).

In Europe, while consequences of climate change for agriculture will vary widely depending on location and farming system, in general, northern Europe is expected to benefit from increases in yield and climatically suitable areas for cultivation, whereas southern Europe will be disadvantaged by decreased water availability and extreme weather events (Bindi and Olesen, 2011). According to modelling done by Ewert et al. (2005) based on Intergovernmental Panel on Climate Change (IPCC) emission scenarios, the combined effects of technology, climate and CO_2 concentrations could lead to an increase in European crop productivity of up to 163 per cent (compared to 2000) by 2080, although technology was by far the most important driver of this. Such productivity changes could potentially lead to agricultural abandonment in Europe, as productivity is expected to exceed demand (Ewert et al., 2005). On the other hand, climate variability is largely responsible for variations in crop productivity and the variability of crop yields has increased since the beginning of the twenty-first century due to extreme weather events, which are expected to become more frequent in future (Lavalle et al., 2009). Other changes in land use and cropping are also likely. For instance, energy crops and crops currently grown mainly in the south (e.g. maize, sunflowers and soybeans) may shift northwards (Bindi and Olesen, 2011). As ever, however, the picture is a complex one and, as demonstrated in our discussion of the UK context below, all regions and farm types are likely to experience both positive and negative effects and will need to adapt their management practices in order to minimise the risks and maximise the opportunities.

We have already started to see changes in our climate, with impacts beginning to materialise. According to the Living with Environmental Change partnership (now the Research and Innovation for our Dynamic Environment (RIDE) Forum[1]), in the UK average temperatures have increased by between 0.8 and 1°C since the 1980s, with the top ten warmest years on record[2] all having occurred since 1990 (Morison and Matthews, 2016). This has brought some benefit to agriculture, as over the period 2006–15 the growing season was, on average, 29 days longer than in 1961–90.[3] Between 2005 and 2014 there were also 15 per cent fewer days of air frost, and 12 per cent fewer days of ground frost compared to 1961–90 (ibid.). On the other hand, flooding has increased in magnitude and frequency over the last 30 years, especially in the north and west, and sea level around the UK rose an average of 1–2 mm per year in the twenty-first century increasing to over 3 mm per year over the past decade, which has already affected estuaries and increased the risk of storm surges. Furthermore, peak ground-level concentrations of ozone, which can affect some sensitive crops, have declined by about 30 per cent in recent decades (ibid.).

In terms of future climate, temperatures in the UK are expected to rise, particularly in the south (by up to 4°C in summer by the 2080s), resulting in a longer growing season and fewer days of frost. However, the accompanying lower summer rainfall, again particularly in the south of the country, is likely to increase the risk of droughts. Increased winter rainfall and continued rises in sea level (and potentially heavier summer rainstorms) will also increase the risk of flooding, although the scale of this increase is unclear (Morison and Matthews, 2016). All this presents a variety of challenges (as well as opportunities) for agriculture and food production, as summarised in Table 2.1.

These are, of course, just a few of the potential impacts on UK agriculture and the precise consequences for productivity in the industry are difficult to predict: not least because impacts will depend on the extent and nature of each climatic change, as well as on the way in which farmers respond (or not) in adapting their management practices. Complex interactions between changing climatic variables will also have indirect impacts on agriculture through affecting its supporting services, including insect pollination, soil function and water and nutrient cycling but, again, the nature of such impacts are uncertain.

Given this wide array of impacts, it is imperative that the agricultural industry take steps to adapt to climate change if food production is to keep up with demand and if individual farm businesses are to survive. Adaptation responses are required at an individual farm business, wider industry and government levels. In the UK context, these might include:

- Developing new crop varieties and livestock breeds better suited to future climatic conditions and resilient to pests and diseases;
- Adapting crop and livestock management to benefit from warmer temperatures (e.g. changing sowing, harvesting, lambing and calving dates);
- Increasing biosecurity provisions and increasing ventilation in livestock housing;
- Improving soil, water and crop management to reduce erosion, flooding, compaction and losses of soil organic carbon and to improve the efficiency of water and fertiliser use;
- Improving water supply for agriculture and increasing capacity for water capture and storage;
- Increasing diversity in farm businesses to build adaptive capacity, as well as resilience to extreme weather events and economic shocks;
- Developing an 'adaptive management' approach that comprises a continuous cycle of planning, implementing and reviewing management changes.

(Morison and Matthews, 2016, p.22).

Table 2.1 Impacts of climate change on UK agriculture

Expected change	*Potential impact on agriculture*	*Level of scientific confidence*
Increase in atmospheric CO_2 concentrations	• Boost to plant productivity (though precise response also depends on other factors).	M
	• Plant water use is decreased under high CO_2 conditions, potentially offsetting negative impact of drier conditions.	H
	• Varying effects on the chemical composition, and therefore nutritional quality of plants.	H
Increased average temperatures	• Longer growing seasons.	H
	• Increased growth, and therefore higher yield potential, for root crops, leafy vegetables and some perennials.	H
	• For some crops, incl. many cereals, warmer temperatures may hasten maturity, reducing crop duration and yield.	H
	• Northward shift of crops such as forage maize and wider uptake of new crops (e.g. grapevines) and varieties.	H
	• Increase in altitude limits for the growth of different crop and tree species.	H
	• Increased occurrence of pests and diseases.	H
	• Adverse effects on yields of current varieties of perennial fruit crops due to receiving insufficient chilling for optimum development.	H
	• Increased likelihood of heat stress to animals (particularly dairy farming in the south).	M
Drier summers	• Reduction in crop and pasture growth, and consequent reduction in yields and livestock productivity.	H
	• Increased demand for irrigation, particularly high-value horticultural crops and potatoes.	H L M
	• Reduced groundwater recharge and increased competition from other water uses may limit the ability to irrigate.	
	• This may lead to shifts in where some crops are grown (e.g. rain-fed potato production may move west).	
Increased winter rainfall	• Increased loss of topsoil and nutrients due to erosion.	M
	• Increased waterlogging of soils, with adverse effects on crop and pasture growth.	M
	• Potential opportunities for storing excess winter rainfall for use during the summer.	

(*continued*)

Table 2.1 (Cont.)

Expected change	*Potential impact on agriculture*	*Level of scientific confidence*
Increased frequency and intensity of extreme weather	• Negative impact on crop growth due to prolonged periods of drought.	H
	• Periods of drought may reduce some pests such as slugs (although these will benefit from milder, wetter winters).	H
	• Adverse impacts on crops and pasture from increased flooding.	M
	• More frequent heatwaves will have negative effects on crop yield and quality, as well as dairy and livestock productivity.	H (crops) M (livestock)
Sea-level rise	• Loss of productive coastal land and increased risk of flooding and salinisation.	H

Source: Based on Morrison and Matthews, 2016

Ultimately, the consequences of climate change for agriculture depend greatly on the capacity of the industry to adapt, and this will vary from country to country and farm to farm. Campbell and Thornton (2014, p.3) suggest that

> the climate resilience of a farmer could be defined by the degree to which he/she can anticipate, endure and recover from the unforeseen shocks of climate-related hazards, but also by their capacity to adapt and become less vulnerable and therefore more resilient.

Smallholder farmers are, and will continue to be, particularly vulnerable, as they often lack the resources and access to education, innovation and finance available to larger enterprises (Campbell and Thornton, 2014). These farmers will thus need particular support in adapting to climate change impacts – both globally and in the UK. Such support might include efforts to reduce market volatility and improve access to financial support such as risk management and insurance schemes, credit and resources (Hart et al., 2017). In general, resilience needs to be built across farm businesses to help protect them from external impacts (including, but not limited to, climate change) and, as Iglesias et al. (2012) point out, adaptation strategies cannot be considered in isolation from wider influences on farmer behaviour such as changes in market prices and the broader socio-economic climate.

As one of the main contributors to greenhouse gas emissions, agriculture is also under pressure to implement climate change mitigation strategies (as well as to mitigate environmental impacts such as soil erosion and diffuse water pollution, which could be exacerbated by climate change), with further implications for management and recording practices at the farm level.

Although there are no clear greenhouse gas emission reduction targets for the agricultural sector, the EU is committed (under the 2016 Paris Agreement) to reducing emissions by at least 40 per cent[4] by 2030 and the UK government is committed (under the UK Climate Change Act 2008) to reducing emissions by at least 80 per cent[5] by 2050: agriculture will be expected to play its part in achieving these targets (Defra, 2017a; Hart et al., 2017), posing further challenges to individual farm businesses.

Responding to the 'perfect storm': from sustainable agriculture to sustainable intensification

According to Pfeffer (1992), the phrase 'sustainable agriculture' was coined first by the Australian agricultural scientist Gordon McClymont, and Wes Jackson is credited with the first publication of the term in 1980. In 1978, Wendell Berry published *The Unsettling of America*, questioning the tenets of industrial agriculture. Jackson (1980) argued that monoculture and high levels of nitrogen fertiliser should be replaced by 'polycultures' of perennials. In 1987, Miguel Altieri coined the term 'agroecology' in his book by the same name. Of course, none of this was new even in the 1980s and the antecedents of sustainable farming can be traced back to the rise of organic farming philosophies and practices in Europe to the inter-war years and before (Conford, 2001). But the notion of sustainability gave some more rigour and analytic power to the organic movement, although arguably in some sense they switched attention away from the social and ethical dimensions that have long been strong in the organic movement to focus much more squarely on environmental concerns.

The rise of the discourse of sustainable agriculture has to be seen in the context of the broader rise of sustainability, and the emergence of sustainable development as a means of re-engaging environmental concerns with socio-economic challenges. The World Commission on Environment and Development (WCED), better known as the Brundtland Commission, submitted its report to the UN in 1987. In contrast to many earlier environmental reports, the Brundtland Report (WCED, 1987) framed its analysis and recommendations on the social and economic as well as environmental concerns. The report expressed the belief that social equity, economic growth and environmental maintenance are simultaneously possible, thus highlighting the three fundamental components of sustainable development; the environment, the economy, and society, which later became known as the triple bottom line (Du Pisani, 2006). This three-legged approach has been criticised for underplaying the importance of the environment. Dawe and Ryan (2003), for instance, argue that the environment should be given more weight than social and economic factors because, rather than being separate from humanity as the model implies, it is fundamental to its existence (i.e. it is not a leg of the stool but the floor upon which it must stand). This may be so in a generic sense, but we would suggest that in the case of agriculture the mantra of sustainable development has come to be most commonly interpreted as a balancing of economic and environmental interests with little consideration of social well-being. Certainly, most accounts of agricultural sustainability, in a developed world context at least, focus overwhelmingly on the environmental aspect with some reference to economics and very little to social. At the risk of simplification, at the time of the 2008 price spike it could be argued that sustainable agriculture in the UK had three distinctive strands.

First, there was the well-established, but essentially niche, sector of organic farming standing well apart from conventional agriculture as the self-proclaimed exemplar of sustainable agriculture. But despite rapid growth in the 1990s and early 2000s, its market share and land take remained modest, and certainly lagged far behind some European countries. The Member States with the largest areas in 2015 were Spain (almost 2 million ha), Italy (about 1.5 million ha) and Germany (1 million ha), together accounting for around 40 per cent of the EU28 total organic area (European Commission, 2016). By the end of 2015, the proportion of the total agricultural area covered by organic farming stood at nearly 6.2 per cent of the total area (Brzezina et al., 2017). By contrast, in the UK organic farming represents only 2.9 per cent of the total farmed area and has declined by 32 per cent since a peak in 2008 (Defra, 2017c). As in many countries, the organic sector is diverse in its interpretations of sustainable agriculture, ranging from 'back to the land' micro production oriented to local markets and eschewing mainstream retail to large-scale farm businesses fully incorporated into the mainstream food chain. This incorporation is often referred to conventionalisation (Guthman, 2004) and has been the subject of widespread debate in the literature (Goldberger, 2011; Hall and Mogyorody, 2001; Lockie and Halpin, 2005; Stassart and Jamar, 2008; Sutherland, 2013). It is interesting to note that although the philosophical underpinnings of organic farming are broad and certainly tend to include social (both farm labour and human nutrition arguments) and economic as well as environmental claims, the EU and UK justification and funding for policy support were derived from agri-*environmental* regulations (Centre for Rural Economics Research, 2002; Lampkin et al., 1999).

Second, Integrated Farming Systems (IFS) or Integrated Farm Management (IFM), as championed by LEAF (Linking Environment and Farming), emerged as a middle way between organic and conventional systems seeking to reduce external inputs and reduce externalities. Morris and Winter (1999) (following the British Agrochemicals Association, 1996; Jordan, 1993; Park et al., 1997; Proost and Matteson, 1997), identified the following principles of IFS:

- Crop rotation – to promote soil structure and fertility and reduce agrochemical demand. A minimum of four different crops in rotation is recommended;
- Minimum soil cultivation – this has both agronomic and environmental benefits (e.g. reduced soil erosion and nitrogen volatilisation) and use of mechanical tools for weed control;

- Use of disease resistant cultivars which enable reduced input use;
- Modifications to sowing times – e.g. later sowing times to reduce pest and disease outbreak;
- Targeted application of nutrients – to save costs (by reducing the overall amount of chemical applied) and to benefit the environment (e.g. by reducing chemical contamination of groundwater);
- Rational use of pesticides – e.g. avoidance of prophylactic spraying through crop monitoring and use of thresholds to determine the most appropriate timing of application;
- Management of field margins to create habitats for predators;
- Use of tillage systems that favour natural control of pests, improve soil structure and reduce demand for external nitrogen;
- Modifications to cropping sequences to increase crop diversity;
- Promotion of biodiversity or 'ecological infrastructure management' (3–5 per cent of total cropping area is recommended as non-agricultural vegetation) for ecological benefits and promotion of beneficial predators.

The third key strand of sustainable agriculture thinking to have emerged strongly in the 1990s was farming in which (some of) the environmental negatives associated with the productivist agriculture of the post-war era were tackled through agri-environmental schemes. This is an immense subject, about which we have more to say in Chapter 8 of this book, with much debate about the efficacy of schemes in delivering environmental benefits (Batáry et al., 2015; Jones et al., 2017; Kleijn and Sutherland, 2003; McCracken et al., 2015; Whittingham, 2011) and/or the implications for the behaviour and beliefs of farmers, which has become a rich seam of debate (Burton and Paragahawewa, 2011; Mills, 2012; Riley, 2016; Wilson and Hart, 2000). But here our point is simple – much of the agri-environmental 'turn' has not been about in-field agriculture per se – though of course there are some examples such as the maintenance of over-winter stubbles – but about the setting aside of features, habitats, margins, areas of cropping (for wild bird food) and so forth. The diversion of land and of farmers' energies from agricultural production clearly had an impact on production and self-sufficiency levels but, arguably, did little to promote a notion of holistic sustainable agriculture and certainly not of sustainable intensification, although there are some links with IFS.

But these three farming systems and the narratives associated with them all came under scrutiny and pressure as the policy narrative inevitably changed as a result of the combination of food price increases and the perfect storm analogy. Put crudely, no longer could agricultural sustainability be seen merely as a brake on production, as under the agri-environmental paradigm, nor as a niche consumer space as with organics. IFM, as we see below, was arguably rather better placed to face the new world. A renewed emphasis on production could not be pursued regardless of non-production outcomes. The challenge as it emerged in the late 2000s was quite clearly how to increase

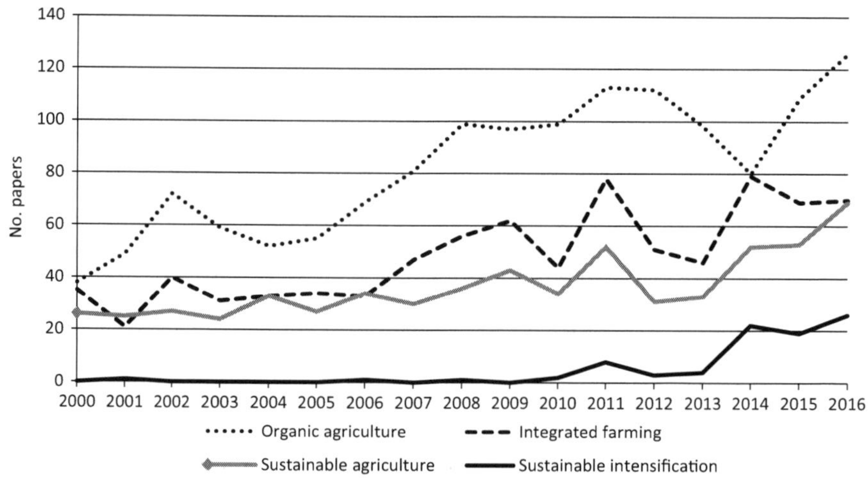

Figure 2.10 Use of terms in the titles of Web of Science papers, 2000–2016

production in a manner both environmentally and socially benign. The term coined to describe this challenge is sustainable intensification (SI). We will provide a brief introduction to SI here and return to it in much more detail in Chapters 5 and 9. SI had its origins in an African context in the mid-1990s (e.g. Pretty, 1997; Reardon et al., 1997) but only emerged as a concept with wide applicability and scientific use a decade or more later; as shown in Figure 2.10, which shows the increase in use of the term in the titles of science papers, though note it still lags behind use of other terms.

Smith (2013, p.19) defines SI as 'the process of delivering more safe, nutritious food per unit of input resource, whilst allowing the current generation to meet its needs without compromising the ability of future generations to meet their own needs'. So rather than simply producing food that does not produce negative externalities or closes the yield gap (Mueller et al., 2012), this definition of SI implies there is a requirement to maintain the long-term capacity of land to produce food and other services (Garnett et al., 2013; Gunton et al., 2016; Neufeldt et al., 2013). This means protecting ecological functions, a different vision of sustainability than simply reducing agro-chemical or faecal pollution or soil erosion through greater efficiency in agricultural production systems, and also including social and ethical considerations surrounding labour rights, animal welfare and gender equality as well as agro-ecological parameters (Gunton et al., 2015). As Garnett and Godfray (2012, p.43) have argued, 'sustainable intensification, if it is to be a meaningful aspiration, needs to be mindful of the social, economic and ethical context within which food production activities take place'. And the place of SI in a wider food security debate remains not entirely clear. According to Cook et al. (2015), SI does not provide a paradigm for achieving food security overall, but is important as one element of a wider sustainable food system. A concluding point, given the focus of this book, is that SI does not necessarily have a better answer to the social question in sustainable agriculture than other approaches; an issue to which we will return.

Conclusion

Our overview of supply-side challenges and the introduction of SI as a possible solution to these challenges does not exhaust the issues that might be relevant to agriculture in the era of food security. A commonly cited alternative to increasing food production is through the reduction of food waste. In developing countries, food wastage primarily occurs during the production, transport and processing stages, whereas in developed countries home and municipal waste is the single largest factor (Godfray et al., 2010). In the UK, annual food waste post-farm gate is estimated at around 10 million tonnes (excluding wastage occurring outside of the UK in the production and transport of food imports), which is around a quarter of all food purchased (WRAP, 2017). But reducing this is hugely challenging – a truly 'wicked problem' – requiring multiple behavioural, regulatory and technical changes across the diversity of actors in a complex food system. The same, of course, also applies to eating habits as others have advocated reduced consumption, particularly of meat, as a means of reducing demand and responding to the food security challenge (Weis, 2013). The waste and diet issues have not been given as strong an emphasis in this book as perhaps they warrant because our concern, in this text, is primarily with farming as it is practised.

Nor have we given much attention in this chapter to the 'social question' in agriculture which we signalled in the first chapter as one of our main concerns. This has been a deliberate omission. Why? Because the supply/demand metaphor which has framed this chapter is the conventional orthodoxy for reviewing these issues. At one level this is quite right. Agricultural production is about the deployment of physical inputs to a physical resource, land. But the deployment (the farming) and the land management are undertaken by farmers, farm workers, contractors, agronomists, and so forth. Calling them 'human capital', not a term we much like, makes the point that people matter in the economics of agricultural production. And we don't like the capital metaphor any more than we like the notion of 'human resources'. Our interest, and our concern, is for people – thinking, feeling, hurting, resilient people.

Notes

1 RIDE is 'a forum of 19 public sector member organisations who hold a stake in environmental change research, innovation, training and capabilities, whether that be as funders, providers and/or users'. www.nerc.ac.uk/research/partnerships/ride/.

2 Records began in 1910.
3 Calculated with the Central England Temperature record as the period where daily mean temperature is above 5°C for more than five consecutive days.
4 From 1990 levels.
5 From 1990 levels.

References

Altieri, M. (1987). *Agroecology: The scientific basis of alternative agriculture.* Florida: CRC Press.

Badgley, C., Moghtader, J., Quintero, E., Zakem, E., Chappell, M. J., Avilés-Vázquez, K., Samulon, A. & Perfecto, I. (2007). Organic agriculture and the global food supply. *Renewable Agriculture and Food Systems*, 22(2), 86–108.

Baines, J. (2015). Fuel, feed and the corporate restructuring of the food regime. *The Journal of Peasant Studies*, 42, 295–321.

Batáry, P., Dicks, L. V., Kleijn, D. & Sutherland, W. J. (2015). The role of agri-environment schemes in conservation and environmental management. *Conservation Biology*, 29(4), 1006–1016.

Baumeister, C. & Kilian, L. (2014). Do oil price increases cause higher food prices? *Economic Policy*, 29(80), 691–747.

Beddington, J. (2009). *Food, Energy, Water and the Climate: A perfect storm of global events?* London: Government Office for Science.

Berti, P. R., Desrochers, R. E., Van, H. P., Văn, A. L., Ngo, T. D., The, K. H., Le Thi, N. & Wangpakapattanawong, P. (2016). The process of developing a nutrition-sensitive agriculture intervention: A multi-site experience. *Food Security*, 8(6), 1053–1068. Available at: http://link.springer.com/10.1007/s12571-016-0625-3 [accessed 15/08/18].

Bindi, M. & Olesen, J. E. (2011). The responses of agriculture in Europe to climate change. *Regional Environmental Change*, 11(1), 151–158.

Bowley, H. E., Mathers, A. W., Young, S. D., Macdonald, A. J., Ander, E. L., Watts, M. J., Zhao, F. J., McGrath, S. P., Crout, N. M. J. & Bailey, E. H. (2017). Historical trends in iodine and selenium in soil and herbage at the Park Grass Experiment, Rothamsted Research, UK. *Soil Use and Management*, 33(2), 252–262.

BP (2017). *Statistical Review of World Energy – Underpinning data, 1965–2016.* [Online] Available at: www.bp.com/en/global/corporate/energy-economics/statistical-review-of-world-energy/downloads.html [accessed 13/03/18].

British Agrochemicals Association (1996). *Integrated Crop Management.* Peterborough: BAA.

Brzezina, N., Biely, K., Helfgott, A., Kopainsky, B., Vervoort, J. & Mathijs, E. (2017). Development of organic farming in Europe at the crossroads: Looking for the way forward through system archetypes lenses. *Sustainability*, 9(5), 821.

Burton, R. J. F. & Paragahawewa, U. H. (2011). Creating culturally sustainable agri-environmental schemes. *Journal of Rural Studies*, 27(1), 95–104.

Campbell, B. & Thornton, P. (2014). *How Many Farmers In 2030 And How Many Will Adopt Climate Resilient Innovations? CCAFS Info Note.* Copenhagen: CGIAR Research Program on Climate Change, Agriculture and Food Security (CCAFS).

Centre for Rural Economics Research (2002). *Economic Evaluation of the Organic Farming Scheme. Final report to Defra.* Cambridge: University of Cambridge.

Chamberlain, D. E., Fuller, R. J., Bunce, R. G. H., Duckworth, J. C. & Shrubb, M. (2000). Changes in the abundance of farmland birds in relation to the timing of agricultural intensification in England and Wales. *Journal of Applied Ecology*, 37(5), 771–788.

Conford, P. (2001). *The Origins of the Organic Movement.* Edinburgh: Floris Books.

Cook, S., Silici, L., Adolph, B. & Walker, S. (2015). *Sustainable Intensification Revisited.* London: IEED Issues Paper.

Cooper, J. & Carliell-Marquet, C. (2013). A substance flow analysis of phosphorus in the UK food production and consumption system. *Resources, Conservation and Recycling*, 74, 82–100.

Cordell, D., Drangert, J.-O. & White, S. (2009). The story of phosphorus: Global food security and food for thought. *Global Environmental Change*, 19(2), 292–305.

Corstanje, R., Deeks, L. R., Whitmore, A. P., Gregory, A. S. & Ritz, K. (2015). Probing the basis of soil resilience. *Soil Use and Management*, 31(S1), 72–81.

Dawe, N. K. & Ryan, K. L. (2003). The faulty three-legged-stool model of sustainable development. *Conservation Biology*, 17(5), 1458–1460.

Dawson, C. J. & Hilton, J. (2011). Fertiliser availability in a resource-limited world: Production and recycling of nitrogen and phosphorus. *Food Policy*, 36, S14–S22.

Defra (2014). *A series showing the UK's food production to supply ratio (commonly referred to as the 'self-sufficiency' ratio) from 1956*. [Online] Available at: www.gov.uk/government/statistical-data-sets/overseas-trade-in-food-feed-and-drink [accessed 07/03/18].

Defra (2017a). *Agricultural Statistics and Climate Change. Eighth edition.* London: Defra.

Defra (2017b). *Food Statistics Pocketbook 2017: Data for chart 3.2 UK food production to supply ration 1998 to 2016*. [Online] Available at: www.gov.uk/government/statistics/food-statistics-pocketbook-2017 [accessed 07/03/18].

Defra (2017c). *Organic Farming Statistics 2016*. London: Defra.

Dobbs, R., Sawers, C., Thompson, F., Manyika, J., Woetzel, J., Child, P., McKenna, S. & Spatharou, A. (2014). *Overcoming Obesity: An initial economic analysis.* McKinsey Global Institute.

Du Pisani, J. A. (2006). Sustainable development – historical roots of the concept. *Environmental Sciences*, 3(2), 83–96.

Erisman, J. W., Sutton, M.A., Kilmont, J., Galloway, Z. & Winiwarter, W. (2008) How a century of ammonia synthesis has changed the world. *Nature Geoscience*, 1(6), 636–639.

European Commission (2016). *Facts and Figures on Organic Agriculture in the European Union 2016*. Brussels: European Commission

Evans, N., Morris, C. & Winter, M. (2002). Conceptualizing agriculture: A critique of post-productivism as the new orthodoxy. *Progress in Human Geography*, 26(3), 313–332.

Ewert, F., Rounsevell, M. D. A., Reginster, I., Metzger, M. J. & Leemans, R. (2005). Future scenarios of European agricultural land use: I. Estimating changes in crop productivity. *Agriculture, Ecosystems & Environment*, 107(2), 101–116.

Farnworth, G. & Melchett, P. (2015). *Runaway Maize*. Bristol: Soil Association.

Fish, R., Lobley, M. & Winter, M. (2013). A license to produce? Farmer interpretations of the new food security agenda. *Journal of Rural Studies*, 29, 40–49.

Flynn, A. (1986). Agricultural policy and party politics in post-war Britain. *In:* Cox, G., Lowe, P. & Winter, M. (eds) *Agriculture: People and policies*. London: Allen and Unwin.

Food and Agriculture Organization (2013). *The State of Food and Agriculture: Food systems for better nutrition*. Rome: FAO.

Food and Agriculture Organization (2015). *FAO Statistical Pocketbook: World food and agriculture 2015*. Rome: FAO.

Food and Agriculture Organization (2016). *The State of Food and Agriculture: Climate change, agriculture and food security*. Rome: FAO.

Food and Agriculture Organization. (2017). *FAO Food Price Index* [Online]. Available at: www.fao.org/worldfoodsituation/foodpricesindex/en/ [accessed 17/10/17].

Food and Agriculture Organization (2018). *FAO Food Price Index in nominal and real terms*. [Online] Available at: www.fao.org/worldfoodsituation/foodpricesindex/en/ [accessed 07.03.18].

Foster, C., Green, K., Bleda, M., Dewick, P., Evans, B., Flynn, A. & Mylan, J. (2006). *Environmental Impacts of Food Production and Consumption: A report to the Department for Environment, Food and Rural Affairs*. Manchester Business School, The University of Manchester.

Freitag, S., Verrall, S. R., Pont, S. D. A., McRae, D., Sungurtas, J. A., Palau, R., Hawes, C., Alexander, C. J., Allwood, J. W., Foito, A., Stewart, D. & Shepherd, L. V. T. (2018). Impact of conventional and integrated management systems on the water-soluble vitamin content in potatoes, field beans, and cereals. *Journal of Agricultural and Food Chemistry*, 66(4), 831–841.

Garnett, T., Appleby, M. C., Balmford, A., Bateman, I. J., Benton, T. G., Bloomer, P., Burlingame, B., Dawkins, M., Dolan, L., Fraser, D., Herrero, M., Hoffmann, I., Smith, P., Thornton, P. K., Toulmin, C., Vermeulen, S. J. & Godfray, H. C. J. (2013). Sustainable intensification in agriculture: Premises and policies. *Science*, 341(6141), 33–34.

Garnett, T. & Godfray, C. (2012). *Sustainable Intensification in Agriculture. Navigating a course through competing food system priorities*. University of Oxford: Food Climate Research Network and the Oxford Martin Programme on the Future of Food.

Global Burden of Metabolic Risk Factors for Chronic Diseases Collaboration (2014). Cardiovascular disease, chronic kidney disease, and diabetes mortality burden of cardiometabolic risk factors from 1980 to 2010: a comparative risk assessment. *The Lancet Diabetes & Endocrinology*, 2(8), 634–647.

Godfray, H. C. J., Beddington, J. R., Crute, I. R., Haddad, L., Lawrence, D., Muir, J. F., Pretty, J., Robinson, S., Thomas, S. M. & Toulmin, C. (2010). Food security: The challenge of feeding 9 billion people. *Science*, 327(5967), 812–818.

Goldberger, J. R. (2011). Conventionalization, civic engagement, and the sustainability of organic agriculture. *Journal of Rural Studies*, 27(3), 288–296.

Goulding, K. W., Trewavas, A. & Giller, K. E. (2009) Can organic farming feed the world? A contribution to the debate on the ability of organic farming systems to provide sustainable supplies of food. *Proceedings International Fertiliser Society*, 663.

Gregory, A. S., Ritz, K., McGrath, S. P., Quinton, J. N., Goulding, K. W. T., Jones, R. J. A., Harris, J. A., Bol, R., Wallace, P., Pilgrim, E. S. & Whitmore, A. P. (2015). A review of the impacts of degradation threats on soil properties in the UK. *Soil Use and Management*, 31(S1), 1–15.

Griffiths, B. S., Hallett, P. D., Kuan, H. L., Gregory, A. S., Watts, C. W. & Whitmore, A. P. (2008). Functional resilience of soil microbial communities depends on both soil structure and microbial community composition. *Biology and Fertility of Soils*, 44(5), 745–754.

Gunton, R. M., Firbank, L. G., Inman, A. & Winter, M. (2015). *Defining Sustainable Intensification and Developing Metrics with respect to Ecosystem Services for the SIP Research Platform.* Report for Defra project LM0302 Sustainable Intensification Research Platform Project 2: Opportunities and Risks for Farming and the Environment at Landscape Scales.

Gunton, R. M., Firbank, L. G., Inman, A. & Winter, D. M. (2016). How scalable is sustainable intensification? *Nature Plants*, 2(5), 16065.

Guthman, J. (2004). *Agrarian Dreams: The paradox of organic farming in California.* Berkeley: University of California Press.

Hall, A. & Mogyorody, V. (2001). Organic Farmers in Ontario: An examination of the conventionalization argument. *Sociologia Ruralis*, 41(4), 399–322.

Hart, K., Allen, B., Keenleyside, C., Nanni, S., Maréchal, A., Paquel, K., Nesbit, M. & Ziemann, J. (2017). *Research for Agri Committee – The Consequences of Climate Change for EU Agriculture. Follow-up to the COP21 – UN Paris Climate Change Conference*. Brussels: European Parliament.

Hawkes, C., Jewell, J. & Allen, K. (2013). A food policy package for healthy diets and the prevention of obesity and diet-related non-communicable diseases: The NOURISHING framework. *Obesity Reviews*, 14, 159–168.

Heffer, P., Prud'homme, M. P. R., Muirheid, B. & Isherwood, K. F. (2006). Phosphorus Fertilisation: Issues and Outlook. *Proceedings International Fertiliser Society, 586*.

Humphries, R. N. & Brazier, R. E. (2018). Exploring the case for a national-scale soil conservation and soil condition framework for evaluating and reporting on environmental and land use policies. *Soil Use and Management*, 34(1), 134–146.

Iglesias, A., Quiroga, S., Moneo, M. & Garrote, L. (2012). From climate change impacts to the development of adaptation strategies: Challenges for agriculture in Europe. *Climatic Change*, 112(1), 143–168.

International Fertilizer Development Center (2010). *World Phosphate Rock Reserves and Resource*. Muscle Shoals, AL: International Fertilizer Development Center.

Jackson, W. (1980). *New Roots for Agriculture.* Nebraska: University of Nebraska.

Jenssen, T. K. & Kongshaug, G. (2003). Energy consumption and greenhouse gas emissions in fertiliser production. *Proceedings International Fertiliser Society*, 509.

Jones, J. I., Murphy, J. F., Anthony, S. G., Arnold, A., Blackburn, J. H., Duerdoth, C. P., Hawczak, A., Hughes, G. O., Pretty, J. L. & Scarlett, P. M. (2017). Do agri-environment schemes result in improved water quality? *Journal of Applied Ecology*, 54(2), 537–546.

Jordan, V. W. L. (ed.) (1993). *Agriculture: Scientific basis for codes of good agricultural practice,* Luxembourg: Office for Official Publications of the European Communities.

Kleijn, D. & Sutherland, W. J. (2003). How effective are European agri-environment schemes in conserving and promoting biodiversity? *Journal of Applied Ecology*, 40(6), 947–969.

Lambin, E. F. & Meyfroidt, P. (2011). Global land use change, economic globalization, and the looming land scarcity. *Proceedings of the National Academy of Sciences*, 108(9), 3465–3472.

Lampkin, N., Foster, C., Padel, S. & Midmore, P. (1999). *The Policy and Regulatory Environment for Organic Farming in Europe.* Stuttgart: University of Hohenheim.

Lavalle, C., Micale, F., Houston, T. D., Camia, A., Hiederer, R., Lazar, C., Conte, C., Amatulli, G. & Genovese, G. (2009). Climate change in Europe. 3. Impact on agriculture and forestry. A review. *Agronomy for Sustainable Development*, 29(3), 433–446.

Lockie, S. & Halpin, D. (2005). The 'conventionalisation' thesis reconsidered: Structural and ideological transformation of Australian organic agriculture. *Sociologia Ruralis*, 45(4), 284–307.

McCracken, M. E., Woodcock, B. A., Lobley, M., Pywell, R. F., Saratsi, E., Swetnam, R. D., Mortimer, S. R., Harris, S. J., Winter, M., Hinsley, S. & Bullock, J. M. (2015). Social and ecological drivers of success in agri-environment schemes: The roles of farmers and environmental context. *Journal of Applied Ecology*, 52(3), 696–705.

Mills, J. (2012). Exploring the social benefits of agri-environment schemes in England. *Journal of Rural Studies*, 28(4), 612–621.

Montague-Fuller, A. (2014). *The Best Use of Agricultural Land.* Cambridge: University of Cambridge Institute for Sustainability Leadership

Morison, J. I. L. & Matthews, R. B. (eds) (2016). *Agriculture and Forestry Climate Change Impacts Summary Report*, Swindon: Living With Environmental Change Network.

Morris, C. & Winter, M. (1999). Integrated farming systems: The third way for European agriculture? *Land Use Policy*, 16(4), 193–205.

Mueller, N. D., Gerber, J. S., Johnston, M., Ray, D. K., Ramankutty, N. & Foley, J. A. (2012). Closing yield gaps through nutrient and water management. *Nature*, 490(7419), 254.

Neufeldt, H., Jahn, M., Campbell, B. M., Beddington, J. R., DeClerck, F., De Pinto, A., Gulledge, J., Hellin, J., Herrero, M. & Jarvis, A. (2013). Beyond climate-smart agriculture: Toward safe operating spaces for global food systems. *Agriculture & Food Security*, 2(1), 12.

Pach, J. D. (2007). Ammonia production: Energy efficiency, CO_2 balances and environmental impact. *Proceedings International Fertiliser Society*, 601.

Palmer, R. C. & Smith, R. P. (2013). Soil structural degradation in SW England and its impact on surface-water runoff generation. *Soil Use and Management*, 29(4), 567–575.

Park, J., Farmer, D. P., Bailey, A. P., Keatinge, J. D. H., Rehman, T. & Tranter, R. B. (1997). Integrated arable farming systems and their potential uptake in the UK. *Farm Management*, 9(10), 483–494.

Pfeffer, M. J. (1992). Sustainable agriculture in historical perspective. *Agriculture and Human Values*, 9(4), 4–11.

Pinstrup-Andersen, P. (2009). Food security: Definition and measurement. *Food Security*, 1(1), 5–7.

Popkin, B. M. (2006). Global nutrition dynamics: The world is shifting rapidly toward a diet linked with noncommunicable diseases. *The American Journal of Clinical Nutrition*, 84(2), 289–298.

Popkin, B. M. (2012). The changing face of global diet and nutrition. *In:* Brownell, K. D. & Gold, M. S. (eds) *Food and Addiction: A comprehensive handbook.* Oxford: Oxford University Press.

Popkin, B. M., Adair, L. S. & Ng, S. W. (2012). Global nutrition transition and the pandemic of obesity in developing countries. *Nutrition Reviews*, 70(1), 3–21.

Popkin, B. M., Horton, S., Kim, S., Mahal, A. & Shuigao, J. (2001). Trends in diet, nutritional status, and diet-related noncommunicable diseases in China and India: The economic costs of the nutrition transition. *Nutrition Reviews*, 59(12), 379–390.

Pretty, J. N. (1997) The sustainable intensification of agriculture. *Natural Resources Forum*, 1997. Wiley Online Library, 247–256.

Proost, J. & Matteson, P. (1997). Integrated farming in the Netherlands: Flirtation or solid change? *Outlook on Agriculture*, 26(2), 87–94.

Reardon, T., Kelly, V., Crawford, E., Diagana, B., Dioné, J., Savadogo, K. & Boughton, D. (1997). Promoting sustainable intensification and productivity growth in Sahel agriculture after macroeconomic policy reform. *Food Policy*, 22(4), 317–327.

Riley, M. (2016). How does longer term participation in agri-environment schemes [re]shape farmers' environmental dispositions and identities? *Land Use Policy*, 52, 62–75.

Rosegrant, M. W. & Msangi, S. (2014). Consensus and contention in the food-versus-fuel debate. *Annual Review of Environment and Resources*, 39(1), 271–294.

Sheail, J. (1997). Scott revisited: Post-war agriculture, planning and the British countryside. *Journal of Rural Studies*, 13(4), 387–398.

Skinner, J., Lewis, K., Bardon, K., Tucker, P., Catt, J. & Chambers, B. (1997). An overview of the environmental impact of agriculture in the UK. *Journal of Environmental Management*, 50(2), 111–128.

Smil, V. (1999). Long-range perspectives on inorganic fertilizers in global agriculture. Paper presented to *1999 Travis P. Hignett Memorial Lecture.* Florence, Alabama,

Smit, A. L., Bindraban, P. S., Schröder, J. J., Conjin, J. G. & van Der Meer, H. G. (2009). *Phosphorus in Agriculture: Global resources, trends and developments*. Wageningen, The Netherlands: Plant Research International B.V.

Smith, P. (2013). Delivering food security without increasing pressure on land. *Global Food Security*, 2(1), 18–23.

Stassart, P. M. & Jamar, D. (2008). Steak up to the horns! The conventionalization of organic stock farming: Knowledge lock-in in the agrifood chain. *GeoJournal*, 73(1), 31–44.

Sutherland, L.-A. (2013). Can organic farmers be 'good farmers'? Adding the 'taste of necessity' to the conventionalization debate. *Agriculture and Human Values*, 30(3), 429–441.

Tadesse, G., Algieri, B., Kalkuhl, M. & von Braun, J. (2014). Drivers and triggers of international food price spikes and volatility. *Food Policy*, 47, 117–128.

The Royal Society (2009). *Reaping the Benefits: Science and the sustainable intensification of global agriculture*. RS Policy document 11/09. London: The Royal Society.

United Nations (2017). *World Population Prospects: The 2017 revision*. [Online] Available at: https://esa.un.org/unpd/wpp/ [accessed 15/07/17].

United States Geological Survey (2010). *Phosphate Rock: World Mine Production and Reserves*. Reston, VA: United States Geological Survey.

Vaclav, S. (2004). *Enriching the Earth: Fritz Haber, Carl Bosch, and the Transformation of World Food Production*. Cambridge, MA and London: MIT Press.

WCED (1987). *Our Common Future*. Oxford: Oxford University Press.

Weis, T. (2013). The meat of the global food crisis. *The Journal of Peasant Studies*, 40(1), 65–85.

Whittingham, M. J. (2011). The future of agri-environment schemes: Biodiversity gains and ecosystem service delivery? *Journal of Applied Ecology*, 48(3), 509–513.

Wiggins, S., Keats, S. & Compton, J. (2010). *What Caused the Food Price Spike of 2007/08? Lessons for world cereals markets*. London: Overseas Development Institute.

Wilson, G. A. (2001). From productivism to post-productivism … and back again? Exploring the (un)changed natural and mental landscapes of European agriculture. *Transactions of the Institute of British Geographers*, 26(1), 77–102.

Wilson, G. A. & Hart, K. (2000). Financial imperative or conservation concern? EU farmers' motivations for participation in voluntary agri-environmental schemes. *Environment and Planning A*, 32(12), 2161–2185.

Winson, A. (2004). Bringing political economy into the debate on the obesity epidemic. *Agriculture and Human Values*, 21(4), 299–312.

Winson, A. (2013). *The Industrial Diet: The degradation of food and the struggle for healthy eating.* New York: New York University Press.

Winson, A. & Choi, J. Y. (2017). Dietary regimes and the nutrition transition: Bridging disciplinary domains. *Agriculture and Human Values*, 34(3), 559–572.

Winter, M. (1996). *Rural Politics: Policies for agriculture, forestry and the environment.* London: Routledge.

Winter, M. (2018). *Changing Food Cultures: Challenges and opportunities for UK agriculture. Report to the Nuffield Farming Scholarships Trust.*

World Economic Forum (2011). *Global Risks 2011: Sixth edition.* Geneva: World Economic Forum.

World Health Organization (2015). *Global Health Observatory (GHO) data: NCD mortality and morbidity*. [Online] Available at: www.who.int/gho/ncd/mortality_morbidity/en/ [accessed 01/08/17].

WRAP (2017). *Estimates of Food Surplus and Waste Arisings in the UK.* Banbury: WRAP.

Yu, S. & Tian, L. (2018). Breeding major cereal grains through the lens of nutrition sensitivity. *Molecular Plant*, 11(1), 23–30.

3 Farming trends and agricultural restructuring

Introduction

Moving on from the global trends and challenges identified in the previous chapter, we now turn to a discussion about how these play out at the national and local scales through an examination of farming trends and the process of agricultural restructuring. And we begin to bring farming people in.

To the casual observer, the common image of a somewhat bucolic and unchanging English countryside belies the post-war experience of agricultural and rural change. As Oliver Rackham (1986, p.26) famously observed:

> Except for town expansion, almost every hedge, wood, heath, fen, etc. on the Ordnance Survey large scale maps of 1870 is still there on the air photographs of 1940. ... Much of England in 1945 would have been instantly recognisable by Sir Thomas More, and some areas would have been recognised by the Emperor Claudius.

The extent of change and its impact on landscapes and habitats since then has been well documented (Butler et al., 2007; Murdoch and Marsden, 1994; Stoate et al., 2009; Woods, 2005). Agricultural restructuring is clearly implicated in this process and although this is frequently presented in terms of a relentless increase in the size and dominance of the largest farms (the largest 19 per cent of farms in the UK now manage 74 per cent of the land area), in the UK in 2016 there were still many more farms less than 20 hectares in size (101,000 holdings) than there were farms over 100 hectares (41,000 holdings) (Defra et al., 2017). The persistence of small farms reflects both their traditional prevalence and other aspects of agricultural restructuring such as the expansion of diversification, pluri-activity and the number of so-called residential farms (such as 'retirement holdings' and 'lifestyle farms' where non-agricultural income sources play a major role). Small farms have not, therefore, remained unchanged in themselves and have had to evolve and adapt in response to a combination of deep-rooted policy signals, a variety of economic drivers, unexpected events and the aspirations of those who own, occupy and manage farms. In short, there are a wide range of drivers of agricultural change and, in turn, a variety of adjustment responses.

While recent decades have indeed witnessed considerable agricultural change, it is important to distinguish between ephemeral, season-to-season change (e.g. Brassley, 1998), year-to-year changes and adjustments, and the broader process of agricultural restructuring (or structural change) which can result in the reorientation of farm businesses. Restructuring can be thought of in terms of major transformative changes that may be multi-dimensional in nature (Hoggart and Paniagua, 2001). Discussions of agricultural restructuring, however, frequently fail to formally define restructuring or to identify the boundary between mere 'change' and 'restructuring'. Those that do tend to focus principally on size. For instance, Zimmerman and Heckelei (2012, p.577), following Goddard et al. (1993) define restructuring as 'the change of the number of farms in different size classes'. For Ilbery and colleagues (2009, p.1), structural change is associated with 'changes in the number, size and layout of farms', with some of the implications of such changes being subsumed within often contested notions of a shift towards post-productivist agriculture.

The conceptualisation of restructuring as macro-scale transitions in the trajectories of farm businesses from and between a productivist, post-productivist/multifunctional and neo-productivist agriculture are also evident elsewhere (Calleja et al., 2012; Ilbery and Maye, 2010). At the farm level this can involve maintaining agricultural production through enlargement or the redeployment of resources into new agricultural enterprises or diversification, the redeployment of human labour to off-farm work and/or contracting, or survival on a hobby farming basis (Bowler, 2002; Marsden et al., 2002; de Roest et al., 2017). While this approach usefully focuses on the process of restructuring involving the redeployment of resources, it is still unclear where mere change ends and restructuring begins. In a report commissioned by the Land Use Policy Group of the UK's statutory conservation, countryside and environment agencies, Savills (2001, p.12) referred to structural change as 'an amalgam of farmers withdrawing from farming, change in the scale and intensity of farm businesses and change in farmers' reliance on non-agricultural income sources'. Arguably, this approach describes a particular pattern of structural change and is thus focused on the *outcome* of restructuring decisions. Around the same time, Entec (2000) referred to restructuring without explicitly defining the term, although it is clear that for them it involves changes in farm size, labour inputs and upstream and downstream connections.

In our approach to understanding more about the process and nature of contemporary agricultural restructuring, agricultural restructuring is taken to mean recombining or redeploying the resources used in agriculture (i.e. the factors of production: land, labour, capital). This is a broader definition of restructuring than simply change in farm size and encompasses a number of ways of reorientating a business. The concern with reorientation rather than mere change means that this definition focuses concern not on day-to-day management decisions but on what might be conceived of as more enduring trajectories or pathways of change. Clearly, such restructuring can take place at many scales and may produce a variety of outcomes including those encompassed by the Savills definition.

It is against this background that this chapter explores key trends in the structure of English agriculture since the 1980s/1990s. We examine trends in key indicators of the structure and economic health of English agriculture, including farm size, farm type and income. An appreciation of these national trends provides important background and context for our more detailed and localised discussions in later chapters. At the same time, however, as we will discuss, this analysis also raises questions about the ability of 'official' data sets such as Defra's June Survey of Agriculture and Horticulture (hereafter referred to as the June Survey)[1] to provide the type of information needed in order to understand the socio-economics of English farming, as opposed to simply its physical characteristics. With this in mind, we go on to discuss some of the nuances and complexities of restructuring at the farm level which underlie broader patterns of agricultural change, and which draw attention to the role of the individual farmer in initiating and adapting to change.

Farm numbers, types and size

Analysis of Defra June Survey data suggests that the number of farm holdings has declined since the 1980s and 1990s, though this trend has plateaued over the last decade. Several methodological changes in the survey (specifically the exclusion of non-commercial farms and the change from classifying farm types by Standard Gross Margins (SGM) to Standard Outputs (SO) in 2010[2]) preclude making direct detailed comparisons over long time periods but, to give some indication, the total number of farm holdings in England (excluding 'other'/'unclassified' farm types) decreased by nearly a third between 1985 and 2016 from 153,766 (significant holdings only) in 1985 to 105,784 in 2016. It should be noted, though, that some of this apparent change is actually a statistical artefact resulting from changes to the survey methodology and merging of multiple holdings actually farmed as a single unit. The analysis below is split into pre- and post-2010 to take account of these issues and allow comparison within (but not between) the time periods.

Changes to the number of farm holdings have not been uniform across farm types or regions. As Table 3.1 and Figure 3.1 show, the overall number of farms decreased substantially between 1995 and 2009; by 19 per cent nationally and 17 per cent in the South West. Beyond these headline figures, however, we can see wide variations by farm type, from declines of 53 per cent and 44 per cent within the dairy and general cropping sectors respectively to an increase of 11 per cent among cereal farms. The decline in dairy farms is unsurprising in light of well-publicised low and volatile milk prices driving many dairy farms to consolidate and intensify, switch the focus of their enterprise, or go out of business entirely. The decline in general cropping farms is most likely indicative of a long-term trend towards specialisation and consolidation, aided by technological developments (Defra, 2008) and

Table 3.1 Number of holdings in England and South West, 1995–2009

*Farm type**	*England*			*South West*		
	1995	*2009*	*% change*	*1995*	*2009*	*% change*
Cereals	19,960	22,160	+11	2,295	3,085	+34
General cropping	12,909	7,275	− 44	786	648	−18
Horticulture	8,558	5,172	− 40	1,532	1,288	−16
Pigs & poultry	4,693	4,322	− 8	906	980	+8
Dairy	19,632	9,196	− 53	7,198	3,290	−54
Grazing livestock (LFA) †	10,513	11,260	+7	2,253	2,037	−10
Grazing livestock (lowland) †	31,024	24,702	− 20	10,419	8,299	−20
Mixed	11,426	7,397	− 35	2,615	2,034	−22
Other	27,397	27,111	− 1	7,416	7,872	+6
All holdings	**146,112**	**118,595**	**− 19**	**35,420**	**29,533**	**−17**

Note: *All figures based on SGM.

† Note that strictly speaking 1995 and 2009 figures are not directly comparable in these categories because prior to 2004 the categories referred to 'cattle and sheep' only, whereas 'grazing livestock' also includes horses (excluding specialist horse enterprises), goats and deer.

Source: June Survey, Defra, 2018b

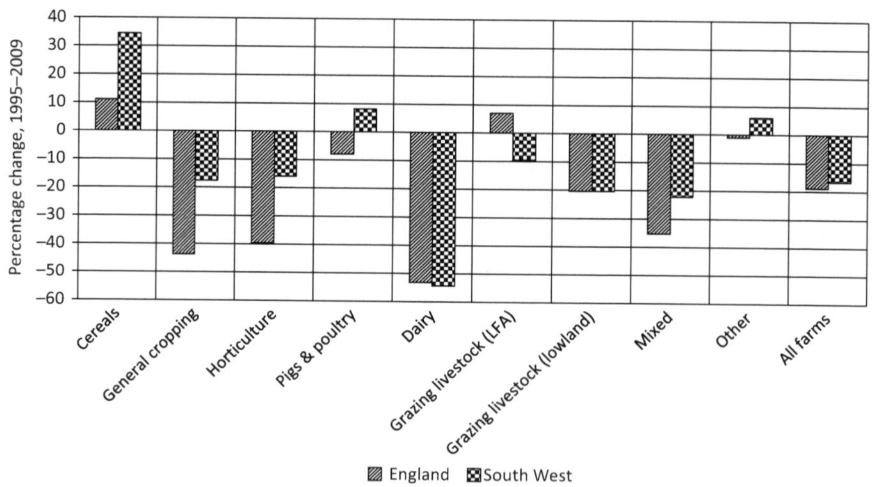

Figure 3.1 Percentage change in the number of farm holdings between 1995 and 2009, England and South West

Source: June Survey, Defra, 2018b

driven by aims to reduce production costs and supply cheap food over the last half-century (de Roest et al., 2017). This may also partly explain the increase in cereal farms. In the South West, while changes to dairy and lowland grazing livestock farm types largely reflect the national picture, there are marked differences elsewhere, with larger increases in the number of cereal farms, lesser decreases in general cropping, horticulture and mixed farms and, in contrast to the national trends, an increase in pig and poultry farms and a decrease in less favoured area (LFA) grazing livestock farms.

Changes to overall farm holding numbers have stabilised somewhat in the period since 2010, although dairy numbers have continued to decline (by 18 per cent) while cereal, pig and poultry farms have increased in number since 2010 (by 14 per cent, 22 per cent and 17 per cent respectively) (see Table 3.2 and Figure 3.2). The pattern of change in the South West has been broadly similar, with the exception of larger increases in the numbers of cereal, pig and mixed farm holdings.

Declines in farm numbers since the 1980s partly reflect an overall reduction in the amount of land used for agriculture across the UK. Competing demands on land for uses such as housing, industry and infrastructure (with consequential high land values), alongside a difficult economic climate for agriculture, has inevitably led to the removal of some land from farming. For instance, between 2006 and 2012, over 4,000 hectares of agricultural land in the UK were lost to construction sites and 5,000 hectares to mineral extraction sites (though nearly 2,000 hectares were converted from mineral extraction sites back to agricultural land) (University of Leicester, 2015). As Figure 3.3 shows, the total area on agricultural holdings in England decreased from around 9.60 million hectares in 1985 to 9.32 million hectares in 2009 (a loss of 284 thousand hectares), though it has remained relatively stable since then at between 8.86 and 9.06 million hectares (using revised figures).[3]

The declining number of holdings is also indicative of a land consolidation process whereby farm sizes are increasing as farmers look to take advantage of new technologies and increase their economies of scale (de Roest et al., 2017; Zimmermann and Heckelei, 2012). Smaller operators have struggled to stay in business in the context of a tough financial climate and volatile commodity markets. Richards et al. (2013) show how restructuring is also driven by retailers, as UK supermarket oligopoly has left many small farms struggling to contend with the buying power of supermarkets (which can push prices below the cost of production) and the audit burden associated with quality assurance schemes. Another driver of farm size expansion is agricultural policy, particularly in terms of the type of subsidies offered. In a study analysing the stated intentions of farmers in nine European countries under a baseline and 'no-CAP' scenario, Bartolini and Viaggi (2013, p.134) conclude that a 'direct effect of the CAP seems to support the willingness to expand/maintain land size in order to exploit opportunities provided by land-connected payments'. The implications of post-Brexit agricultural policy on future changes in farm size in the UK are intriguing in light of such findings and this is an issue we return to in the concluding chapter.

Table 3.2 Number of holdings in England and South West, 2010–2016

*Farm type**	*England*			*South West*		
	2010	*2016*	*% change*	*2010*	*2016*	*% change*
Cereals	16,837	19,118	+14	1,907	2,420	+27
General cropping	16,673	17,728	+ 6	3,532	4,187	+19
Horticulture	4,602	4,259	− 7	1,041	1,029	−1
Pigs	1,601	1,953	+ 22	244	320	+31
Poultry	2,133	2,495	+17	426	488	+15
Dairy	7,882	6,470	− 18	2,878	2,426	−16
Grazing livestock (LFA)	12,625	12,559	− 1	2,386	2,331	−2
Grazing livestock (lowland)	33,391	32,369	− 3	10,704	10,421	−3
Mixed	8,320	8,833	+6	1,926	2,209	+15
Unclassified	1,385	1,069	− 23	377	260	−31
All holdings	**105,449**	**106,853**	**+1**	**25,421**	**26,091**	**+3**

Note: *All figures based on SO.

Source: June Survey, Defra, 2018b

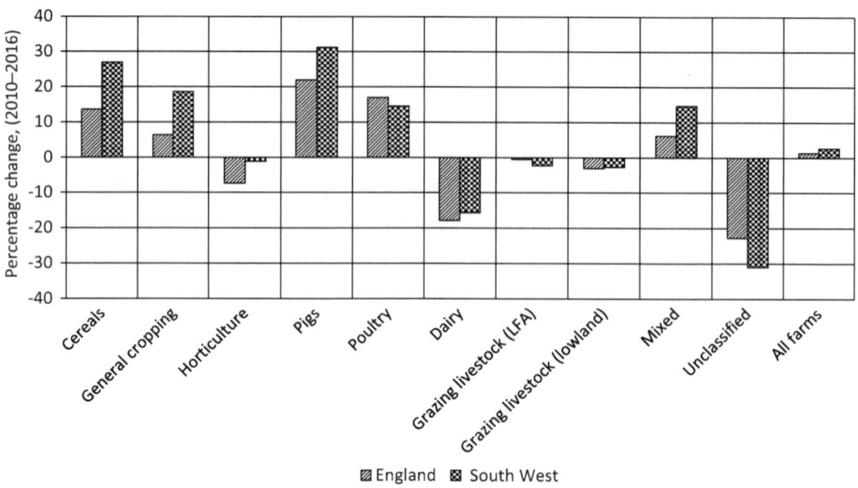

Figure 3.2 Percentage change in the number of farm holdings between 2010 and 2016, England and South West

Source: June Survey, Defra, 2018b

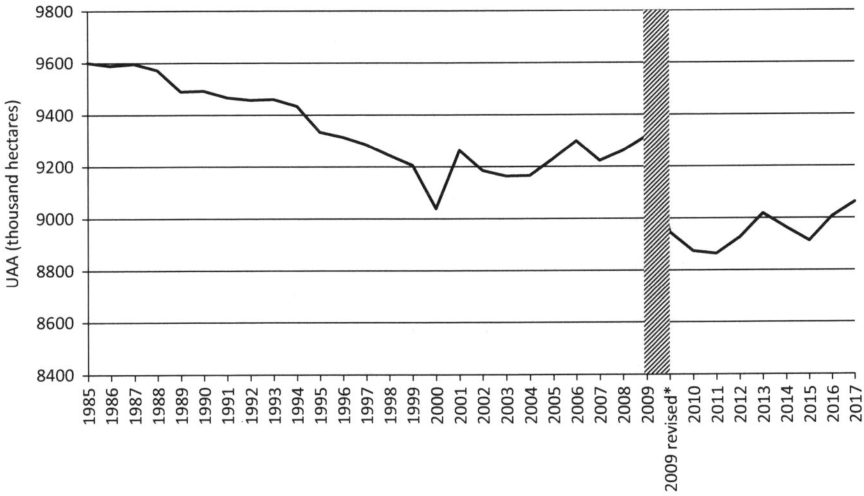

Figure 3.3 Utilised agricultural area in England, 1985–2017

Note: *2009 revised figure (commercial holdings only).

Source: June Survey, Defra, 2018b

For many struggling farm businesses, however, expansion is hindered by a lack of available capital. The value of farmland has increased substantially over recent decades, making farm size expansion a prohibitively expensive option for many operators. For instance, the value of farmland increased 149 per cent over the 10 years to 2017 (varying from 89 per cent in Scotland to 153 per cent in the south-west of England and 210 per cent in the south-east of England) (Savills, 2017b). The average value of prime arable land peaked at almost £10,000 in 2014 (Savills, 2015) before falling slightly to just over £9000 in 2017 (Savills, 2017a). Some of this decline has been related to a fall in commodity prices since 2014 but it is also partly a result of uncertainty in the post-Brexit landscape (RICS Economics, 2017). To some extent, land values decline with agricultural quality (Savills, 2017a), but a number of factors have eroded the correlation between farm profitability and land values over the past century so that the annual returns from agricultural activity on the land are now significantly lower than the capital growth rate (MSCI, 2015). Farmland is an appealing investment for those outside the agricultural sector because it is generally considered to be a 'safe haven' in times of uncertainty and is associated with significant tax benefits in the form of inheritance and capital gains tax reliefs (Savills, 2013). There is also considerable demand for farmland from lifestyle/amenity buyers for residential farms and 'trophy estates' – such buyers accounted for over 40 per cent of all buyers in sales which Savills were involved with in 2017 (Savills, 2018). Thus, the market value of farmland across the developed world is now based less on its productive value and more on what Oltmans (2007) describes as its 'place value' – i.e. the non-farm characteristics and uses it can provide, including recreation, lifestyle enhancement, development potential, investment and speculation, tax benefits, and environmental amenities. Furthermore, the Basic Payment Scheme (BPS) and other agricultural subsidies have a significant influence over land values. Although now dated, an empirical model developed by Harvey (1989) suggested that support policies were resulting in land being overvalued by 46 per cent compared to a free-market model, and that 55 per cent of the gain was being captured by landlords. Changes to agricultural policy and trade following Brexit could, therefore, have a significant impact on UK land values. Savills' (2018) five-year outlook predicts a decline of 2–3 per cent per annum in commercial farmland value due to a weakened trade position and reduced subsidy following Brexit. For land with amenity value where there are options to diversify away from farming, however, the outlook is considerably more positive, with a rise of 3 per cent per annum predicted over the next five years.

Despite the high and arguably distorted price of agricultural land, according to data from the June Survey average farm holding size in England increased by a third from 63.4 to 85.4 hectares between 1995 and 2016. In this period, the number of small farms (5–20 ha) decreased by 25 per cent and the number of very large farms (over 100ha) increased by 22 per cent (see Table 3.3). Similarly, in the South West between 1995 and 2016 average farm size grew by over a third from 51.4 to 68.6 hectares, with the number of small farms decreasing by 24 per cent and the number of very large farms increasing by 32 per cent. As we can see in Table 3.4, 73.6 per cent of all farmed land in England is now held on very large holdings (100 hectares or more) – i.e. by just 24 per cent of all farms. Although comparative detailed data is not available prior to 2010, it is clear from the aforementioned changes in small

Table 3.3 Number of holdings by farm size in England and South West, 1995–2016

	5–20ha		*20–50ha*		*50–100ha*		*>100ha*	
	Eng	*SW*	*Eng*	*SW*	*Eng*	*SW*	*Eng*	*SW*
1995	38,362	10,075	33,756	9,136	24,847	6,372	21,028	3,854
2000	33,543	8,816	27,025	7,111	21,666	5,448	22,007	4,140
2005	38,728	10,308	26,495	7,108	21,531	5,532	26,788	5,213
2010	28,693	7,473	22,244	5,678	19,072	4,798	26,259	5,261
2015	26,522	7,053	20,542	5,376	19,009	4,695	26,202	5,305
2016	28,740	7,635	22,523	5,977	18,388	4,693	25,638	5,095
% change 1995–2016	**−25**	**−24**	**−33**	**−35**	**−26**	**−26**	**+22**	**+32**

Source: June Survey, Defra, 2018b

Table 3.4 Proportion of total farmed area held by farm size category, 2016 (%)

	5<20ha	*20<50ha*	*50<100ha*	*>=100ha*
England	3.5	8.2	14.4	73.6
South West	4.7	11.1	18.7	65.1

Source: June Survey, Defra, 2018b

vs large farm numbers that an increasing amount of land is being held by the largest farms. Figures based on the economic size of farms further reiterate the skewed distribution of agricultural production in England: the largest 7 per cent of farms (those with a SO of at least €500,000) account for 55 per cent (€8.9 billion) of the country's total output, in contrast to the smallest 42 per cent (those with an SO under €25,000), which produce just 2 per cent of total output (Defra and Government Statistical Service, 2018).

Farm labour

Increasing farm size is not commensurate with more labour on farms for a number of reasons, including economies of scale, technological advancements, greater use of contractors and financial pressures. Despite larger farms, total labour has declined from an average of 3.51 persons per farm in 1985 to 2.82 in 2016 (see Table 3.5), equating to a total reduction of 174,221 people working on farms across England. As shown in Figure 3.4, the greatest change has been to the number of full-time workers, which has more than halved from 131,825 in 1985 to 57,896 in 2016. Numbers of regular part-time and casual workers have also reduced substantially.

South West data for 1985 is not available, but average labour on farms reduced from 2.74 persons per farm in 1995 to 2.45 in 2016. The greatest change here has been to the number of casual workers, which has nearly halved from 10,379 in 1995 to 5,415 in 2016, though numbers of regular workers (both full and part-time) have also decreased significantly (see Figure 3.5).

While technological advances are likely to be driving some of this reduction in labour by enabling farmers to cultivate larger areas more efficiently, data from the 2016 SW Farm Survey suggest that (in this region at least) there are more factors at play. Although 22.9 per cent of respondents stated that "changes in technology mean I need less labour on my farm", 56.1 per cent said they needed just as much labour and 6.6 per cent said they needed more. This may be partly related to the relatively low uptake of some recent technologies in the South West: 91.5 per cent of farmers said that they do not currently use any of the technologies listed (robotic tractors, drones, remote monitoring, GPS auto steer) and 91 per cent do not have any plans to do so over the next five years. While directly comparable data is not available at the national level, a Defra survey of farm practices in 2012 found that 22 per cent of farms used

Table 3.5 Average number of workers per farm in England, 1985–2016 and South West, 1995–2016 (persons per farm)

	Farmers, directors & spouses (FT & PT)		*Salaried managers & regular workers (FT)*		*Regular workers (PT)*		*Casual workers*		*Total labour*	
	Eng	*SW*	*Eng*	*SW*	*Eng*	*SW*	*Eng*	*SW*	*Eng*	*SW*
1985	1.5	–	0.9	–	0.3	–	0.5	–	3.1	–
1995	1.7	1.6	0.7	0.5	0.3	0.3	0.5	0.3	3.2	2.7
2016	1.6	1.6	0.5	0.4	0.2	0.2	0.4	0.2	2.8	2.4

Source: June Survey, Defra, 2018b

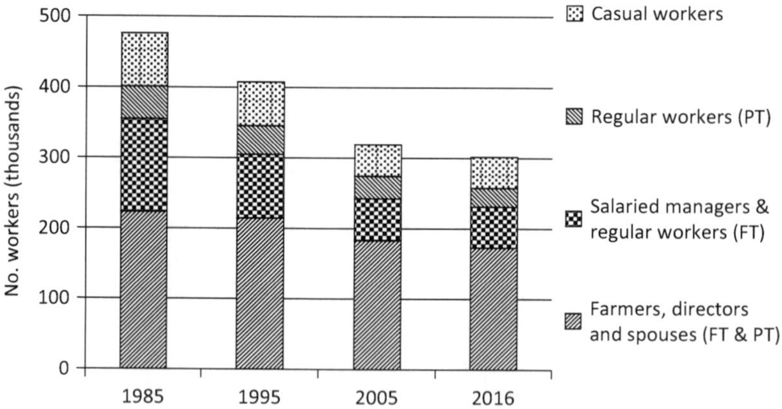

Figure 3.4 Farm labour in England, 1985–2016

Source: June Survey, Defra, 2018b

GPS auto steer (Defra, 2013), indicating that levels of uptake of such technologies is indeed lower in the South West than elsewhere in the country. This is perhaps unsurprising, however, given the limitations associated with comparatively smaller farm sizes and challenging topology present in the region.

Fewer numbers of (particularly casual) workers on farms are conceivably linked to an increasing use of contractors for specific, often seasonal, tasks. Where traditionally most, if not all, of farm work would be carried out by family members or locally employed workers, 89.3 per cent of farmers in the 2015 SIP Baseline Survey, and 86.8 per cent of those in the 2016 SW Farm Survey, said that they used contractors to some extent. These findings are comparable to national data quoted in a 2014 Defra report on contracting: according to the 2011 Farm Business Survey (FBS) 80–100 per cent of most farm types used contract work (slightly lower levels were

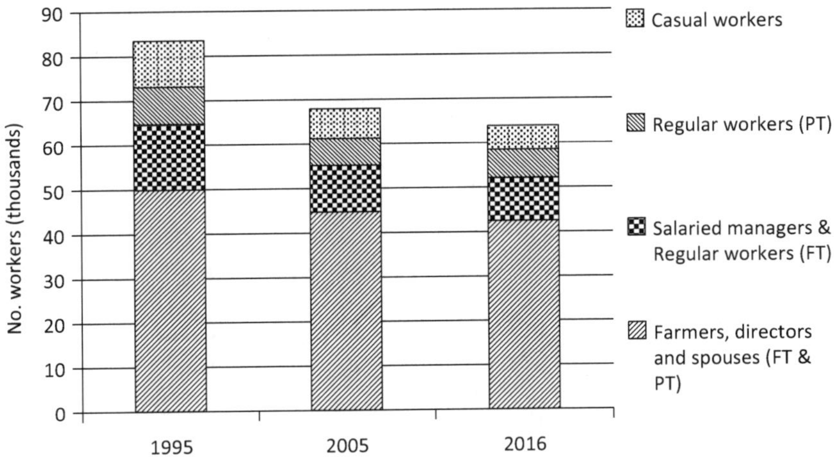

Figure 3.5 Farm labour in the South West, 1995–2016

Source: June Survey, Defra, 2018b

recorded for mixed, horticulture, pigs and poultry farms). Data from the 2010 June Survey show that the use of contractors by most farm types generally increases with size, with the exception of cereal farms where smaller farms have the highest usage (Defra, 2014). Notably, however, contract work generally accounted for a relatively low proportion of total labour requirements (with the exception of very large cereal farms), with most farms still relying on their own family and paid labour for the majority of farm work.

Increasing financial pressures within the agricultural industry are also likely to be influencing the reduction in paid farm employment. Where many large farms may be able to increase efficiency (and reduce labour requirements) by taking advantage of economies of scale and developments in technology, some (particularly smaller) farms are driven to minimise financial expenditure by reducing their number of employed workers and taking on more of the work themselves. As we discuss in Chapter 7, the SW Farm Survey revealed significant numbers of farmers reporting working long hours in an attempt to make their business viable; to the inevitable detriment of their health and well-being.

Further analysis of the 2016 SW Farm Survey and complementary interviews regarding farm labour in the South West was undertaken by Nye (see Nye, 2018; Nye et al., 2017). This research revealed a worrying shortage of skilled and seasonal labour in the region. Thirty-eight percent of surveyed farmers either disagreed or strongly disagreed with the statement 'I can always find skilled labour when required' and this was confirmed by qualitative interviews with farmers, who frequently described job applicants as lacking necessary skills (including basic agricultural skills, as well as those associated with machinery and new technologies), even where those applicants had attended agricultural college. Nye et al. (2017) also report how agricultural contractors in the region have struggled to find suitable employees, with some having to adapt their businesses accordingly. Such labour shortages are particularly concerning in light of the ageing population of UK farmers (as discussed later in this chapter).

Farm business income and economic prospects

Farm businesses face a number of challenges in seeking to ensure a profit from their operations. Concerns over food security in recent years (see Chapter 2) have, in one sense, raised the profile of farmers and proffered hopes of high demand for their products but this has not (yet?) translated into greater financial security within the sector. In today's globalised world, commodity markets are highly volatile with international financial markets, currency movements, competition, consumer demands and worldwide political and climate-related events exerting a strong influence on the price of both inputs and outputs for UK farmers. Local weather conditions also obviously have an impact on agricultural productivity. Thus, average farm business incomes are highly variable from year to year, though as Figure 3.6 shows, longer-term

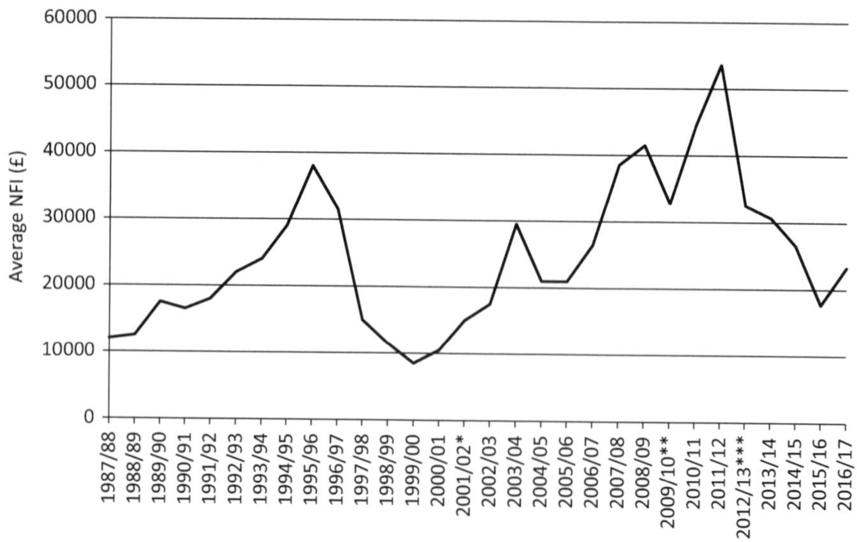

Figure 3.6 Average Net Farm Business Income (NFI), England, 1987–2017

Notes: *Excludes farms subject to foot and mouth disease cull.
**Farm type classification changed from SGM to SO. 2009/10 figure using SGM was £31,500.
***Farm type classification changed from 2007 SO averages to 2010 SO averages. 2012/13 figure using 2007 SO averages was £33,900.

Source: FBS, Defra, 2018a

trends in average net farm business income can also be observed. The significance of currency movements can be illustrated in the impact on the average Basic Payment received by UK farmers, which was 19 per cent higher in 2016 than the previous year, largely due to a weakening of sterling against the euro (Defra et al., 2017).

Although net farm business income (and Farm Business Income (FBI),[4] which is now Defra's preferred measure of income) has fallen sharply in recent years, other measures of performance point to significant variation within the farm sector. For example, analysis of the return on capital employed (ROCE)[5] from the FBS between 2009 and 2017 reveals large disparities between financial performance according to both farm size and type. As Figure 3.7 shows, the larger the farm the higher the average ROCE. Small[6] farms have particularly struggled, on average making a loss in the three years from 2013/14. Medium farms also made an average loss in 2015/16, though recovered slightly in 2016/17.

In terms of farm type, pigs and poultry farms have performed particularly well, with average ROCE peaking at 7.6 per cent in 2013/14 (see Figure 3.8). In contrast, lowland grazing livestock farms have consistently made an average loss since 2012/13, and both LFA grazing livestock and mixed farms have made a loss in four out of the last five years. Market prices account for much of this divergence: for instance, in June 2016 the price spreads from farm to retail for beef and lamb in the UK were 53.5 per cent and 48.3 per cent respectively, compared to the appreciably more rewarding price spreads of 66.9 per cent and 67.9 per cent for pork and bacon respectively (AHDB, 2017d).

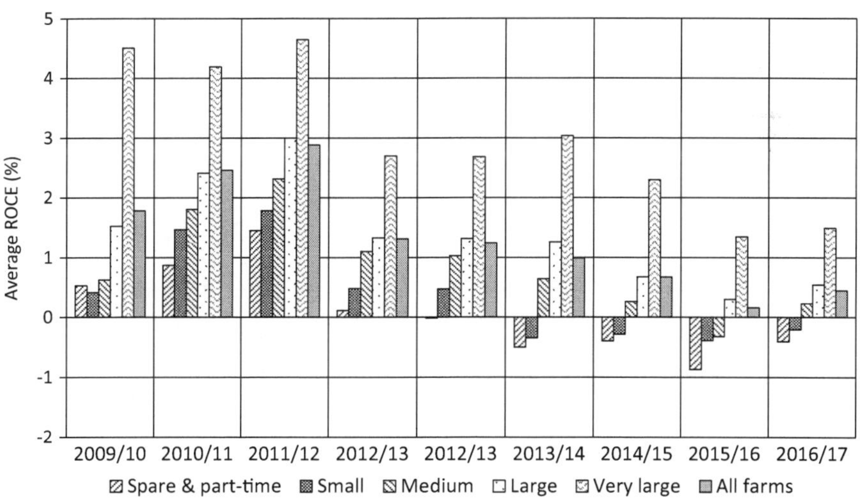

Figure 3.7 Average ROCE by farm size, England, 2009–2017

Source: FBS, Defra, 2018a

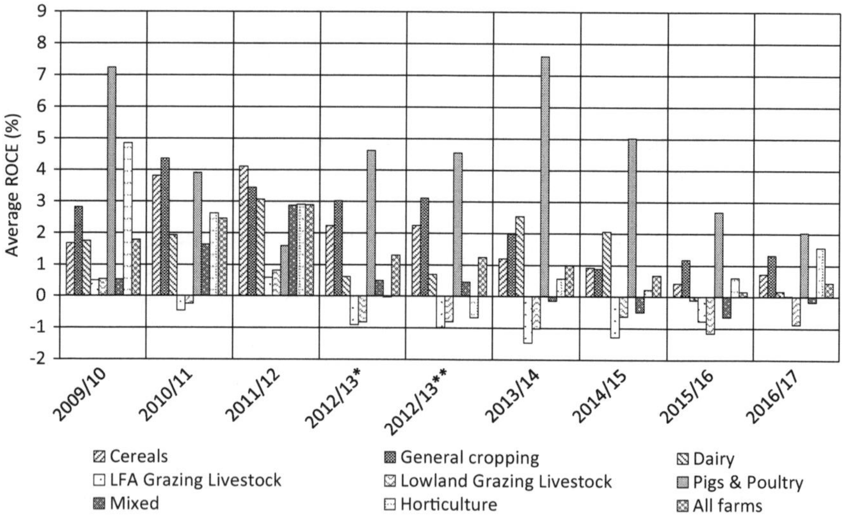

Figure 3.8 Average ROCE by farm type, England, 2009–2017

Notes: *farm typology based on 2007 standard output coefficients.
**farm typology based on 2010 standard output coefficients.

Source: FBS, Defra, 2018a

The story of a challenging economic climate for farm businesses is borne out in data from the SW Farm Surveys. In 2016, 33.5 per cent of respondents perceived their economic prospects over the next five years to be 'poor' or 'bad' and only 15.9 per cent perceived them to be 'good' or 'excellent' (see Table 3.6). Corresponding with fluctuations observed in the national data on farm business incomes, optimism about economic prospects was notably lower than in 2010, although fairly similar to 2006.

The dairy industry is particularly renowned for having faced financial difficulties in recent years and this is an important sector in the South West, which is home to 38 per cent of England's dairy cows and 37 per cent of England's dairy farms and producers, and accounts for 41.5 per cent of England's milk production (Savills and Duchy and Bicton College's Rural Business School, 2017). Recent research by Savills and Duchy and Bicton College's Rural Business School, which looked at 63 dairy farms in the South West between 2010 and 2016, reiterates the precarious nature of the industry. Although financial performance varied considerably between farms with FBI in a 100-cow herd ranging from below £0 to over £100,000 per annum,[7] the farms had an average annual fund deficit (after reinvestment and private drawings) over the six years studied of £16,000, which was financed by increased loans. Ominously, the number of herds in the South West deemed as 'long-term sustainable' under the Agriculture and Horticulture Development Board

Table 3.6 Perceived economic prospects over the next 5 years (% respondents)

	2006	*2010*	*2016*
Excellent	2.2	1.9	1.9
Good	17.7	23.5	14.0
Fair	49.0	54.7	45.7
Poor	23.4	15.5	27.7
Bad	4.9	2.5	5.8

Source: SW Farm Survey

(AHDB) vulnerability index fell from 82 per cent in March 2014 to just 17 per cent in March 2017. Just over half of the remainder are deemed 'short-term sustainable' and the rest 'vulnerable' (Savills and Duchy and Bicton College's Rural Business School, 2017). Fluctuations in milk prices are partly to blame for this situation, placing particular pressure on small farms that cannot absorb short-term losses as well as larger enterprises, but the wide variation in FBI across farms even of a similar size suggest that size is not the only factor in determining success: technology adoption, business acumen and improvements in farm management practices all have a part to play and are not necessarily precluded by small farm size. Nevertheless, dairy farming in the context of today's globalised financial market is not an easy venture. The Russian ban on dairy imports from certain countries including the UK, a weak euro, and high levels of supply together with reduced demand from China have all contributed to the challenges facing the dairy sector, though, more positively, global demand for dairy products is expected to grow by around 2 per cent per annum over the years to 2026 (Downing, 2016).

Productivity and intensification

Technological advances in agriculture, alongside financial drivers to increase efficiency in other ways, have led to a marked increase in the intensity of some operations and in the overall productivity of the industry. As shown in Figure 3.9, trends in the total factor productivity of agriculture[8] show a general improvement over the last 40 years or so, reflecting a combination of both increased outputs and a more efficient use of inputs (including fuel, fertiliser, land and labour). During the 1970s increases in total factor productivity were largely driven by increased outputs, whereas increases between the mid-1990s and mid-2000s were more related to improvements in input efficiency.

Although developments in productivity should be seen positively in terms of this constituting 'one of the main drivers of agricultural income' (Defra and Office for National Statistics, 2017), the relationship is not straightforward. The NFU note that UK growth in productivity is lagging behind that of other developed nations,[9] potentially as a result of underinvestment in research and development (NFU, 2015b) – an important point in the context of a globally

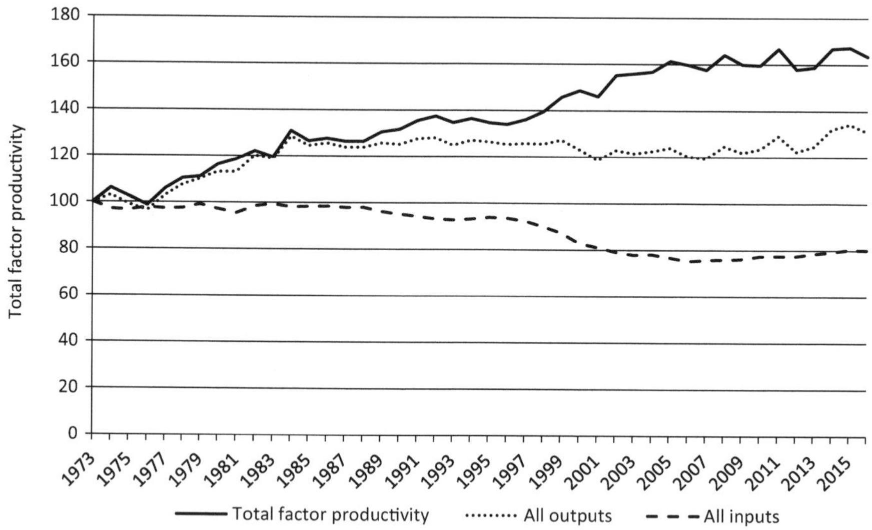

Figure 3.9 Total factor productivity of the UK agricultural industry, 1973–2016

Source: Defra & Office for National Statistics, 2017

competitive market. Productivity improvements must also be considered in terms of structural change and intensification, and all the (potential) environmental costs and animal welfare concerns associated with this. We return to the environmental issue in our discussion of sustainable intensification in Chapter 5, but for now we offer the two examples of dairy and wheat production as illustrations of how the industry has intensified over the last century.

Intensification in the UK dairy industry has been particularly controversial in recent years as stories about 'mega-dairies' with herds of more than 2,000 cows have hit mainstream media headlines. Such farms are perhaps the 'extreme' examples of dairy intensification, but they do reflect a broader tend within the sector towards more intensive production; an intensification facilitated by sophisticated agri-tech developments (e.g. robotic milking systems) and driven by the financial challenges of the dairy industry discussed above. The total number of dairy cows in the UK decreased by 27 per cent over the 20 years from 1996 to 2016, from 2.6 million to 1.9 million cows (see Table 3.7). However, at the same time, average dairy herd size almost doubled from 75 cows in 1996 to 143 in 2016 and average yields increased by around a third from 5,626 litres per cow in 1996 to 7,557 litres per cow in 2016. So, despite the fall in cow numbers, the total amount of milk produced in the UK has remained relatively stable, fluctuating from year to year but reaching a peak of 14 billion litres in 2016/17. The number of dairy farms has also decreased, as noted earlier. Together, these figures indicate a clear intensification of the UK dairy industry over the last two decades. This intensification is reflected

Table 3.7 Dairy herds and milk production in the UK, 1996–2016

	No. cows in the UK (thousand head)	*Average herd size (no. cows)*	*Average yield (litres per cow)*	*Milk production (million litres)**
1996	2,587	75	5,626	13,688
2001	2,251	85	6,449	13,844
2006	1,974	108	6,963	13,481
2011	1,796	123	7,641	13,494
2016	1,897	143	7,557†	14,014

Notes: *UK wholesale deliveries to dairies. Figures are for the financial year beginning in the year stated.
† provisional figure.

Source: AHDB, 2017a, 2017b, 2017c

in figures for the South West. Dairy farms included in the Savills, Duchy and Bicton College's (2017) research increased their utilised agricultural area by an average of just 5 per cent between 2014 and 2017 but their herd size by 16 per cent. Annual milk yield among these farms also rose from 7,631 to 7,990 litres per cow over this time.

As noted in Chapter 2, yields of key crops in the UK have increased significantly and consistently since the Second World War, with wheat seeing the most dramatic rise. In a look back at the last 90 years of wheat production, Savills (2016) report that average wheat yields have increased almost fourfold, from 2.3 tonnes per hectare in 1926 to just under 10 tonnes per hectare in 2015. Yields rose most dramatically in the post-war years, from around 3 tonnes per hectare in the 1940s to around 7 tonnes per hectare in the early 1980s. Of course, increased yields do not necessarily translate directly into increased income for the farm business, however, as the greater supply pushes down prices. Hence, wheat prices fell 84 per cent between 1926 and 2015 from £666 per tonne (real terms) to just over £100 per tonne. This highlights a major challenge facing farmers. Technical change, often embodied in inputs and capital investment, increases the productivity of inputs. The 'agricultural treadmill', a phenomenon initially identified in the USA (Cochrane, 1958, 1979), applies to the UK as well. As the name implies, farmers find themselves in the position of having to 'run' constantly to survive and 'stand still'. Without that capacity to acquire more resources, or innovate, or survive on the basis of subsidising their farming activities from other sources, their farm business is doomed to stagnation at best or extinction at worst. Although UK agriculture is recognised to face a productivity challenge, as we have seen agricultural productivity has improved markedly over recent decades. To an extent, such developments are involuntary for farmers because improvements are embodied in the resources they acquire by purchases such as licensed higher-yielding crop varieties, genetically improved breeding animals, more effective agrochemicals, and machinery technology that facilitates better cultivation practices. However, and a significant point to note in the context of both this chapter and the entire book, the ability to exploit resources to full potential also depends on farmers' own awareness, willingness and ability to implement changes, including technical and business acumen and access to funds for investment.

Sources of income

Low commodity prices and losses on agricultural products mean that, in many cases, the viability of the farm business is heavily dependent on income from the BPS, agri-environment work and diversified activities. The FBS data presented in Figure 3.10 show that in 2016/17 BPS payments made up the largest proportion of farm business income for all farm types except horticulture, pigs and poultry. In total these BPS payments accounted for 66 per cent of all farm income. Agri-environment scheme (AES) payments were also particularly significant for LFA grazing livestock farms, accounting for 42 per cent of total FBI. These payments also accounted for 19 per cent and 15 per cent of lowland grazing and mixed farm income respectively, and for 11 per cent of farm income across all farm types.

Across all farm types, in 2016/17 diversification accounted for 29 per cent of total farm business income across England, and 24 per cent in the South West. The types of diversified activities undertaken by farms are wide and varied but by far the most significant in terms of income is the renting out

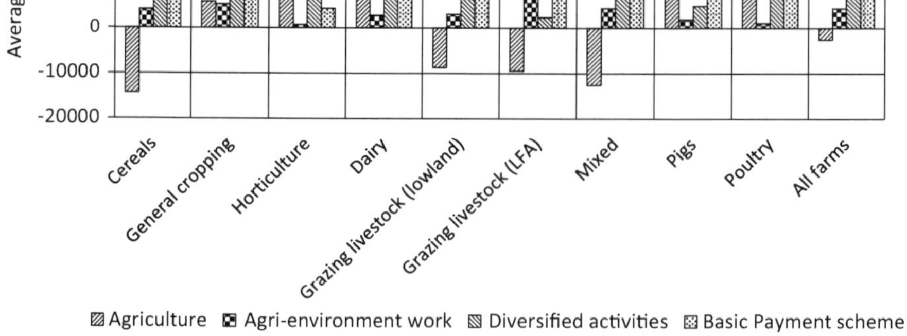

Figure 3.10 Source of Farm Business Income (FBI) by farm type, England, 2016/17

Source: 2016/17 FBS, Defra & Rural Business Research, 2018

of buildings for non-farming use, making up 20 per cent of total FBI (15 per cent in the South West) (see Table 3.8). Other activities appear to make a relatively small contribution to overall FBI but, given the narrow or negative margins received from agriculture, these are not necessarily insignificant in terms of overall farm business or household income. The dependence of individual farms on income from diversified activities is highlighted by a closer look at the contribution of diversification to the total farm income *of those farms which are engaged with each activity* (as opposed to all farms across England/SW, as in Table 3.8). As shown in Table 3.9, this analysis reveals that, on average, farms engaged in diversified activities derive 37 per cent of their income from this source. The letting of buildings for non-farm use and the processing and retailing of farm produce are particularly valuable to those businesses engaged in these activities, contributing 33 per cent and 29 per cent respectively to overall farm income.

Findings from the SW Farm Surveys provide us with additional information about the extent and type of diversified activities undertaken by farmers and how this has changed over the course of a decade (see Table 3.10). In

Table 3.8 Proportion of total farm business income from diversified activities, England and South West, 2015/16 (%)

	England	*South West*
Rent of buildings for non-farming use	20.4	14.9
Processing and retailing of farm produce	2.3	2.0
Sport and recreation	1.2	1.2
Tourist accommodation and catering	1.1	0.8
Other diversified activities	3.9	4.8
All diversified activities	**29.0**	**23.8**

Source: 2016/17 FBS, Defra and Rural Business Research, 2018

Table 3.9 Average contribution from diversification to total income for those farms engaged in the activity, England, 2016/17 (%)

Diversified activity	*Contribution to farm income*
Rent of buildings for non-farming use	33.1
Processing and retailing of farm produce	29.4
Sport and recreation	8.3
Tourist accommodation and catering	13.3
Solar energy	4.8
Other sources of renewable energy	6.7
Other diversified activities	12.0
Diversified enterprises (all kinds)	36.7

Source: 2016/17 FBS, Defra and Rural Business Research, 2018

Table 3.10 Percentage of farms engaged in non-farming enterprises, South West, 2006, 2010 and 2016

	2006	*2010*	*2016*
Non-farming activities (of any type)	**57.0**	**64.4**	**66.6**
Processing and/or retailing of farm produce	6.4	9.2	6.8
Tourist accommodation	14.0	15.0	14.1
Rents from longer-term letting	22.4	–	–
Rents from commercial letting	–	10.1	12.5
Rents from long-term residential letting	–	18.1	24.9
Shooting	8.6	8.2	6.8
Other recreation	2.9	3.4	3.0
Equine services	8.6	7.8	8.3
Forestry	3.8	4.5	5.1
Agricultural services (e.g. contracting)	–	13.0	12.2
Rural crafts	0.3	–	–
Other	9.9	12.6	11.4

Source: SW Farm Survey

2016, two-thirds (66.6 per cent) of all farms surveyed operated a non-farming activity of some sort, up from 64.4 per cent in 2010 and 57 per cent in 2006. The biggest increase has been to the numbers of farms renting out buildings for long-term residential use (which was also the activity undertaken by the largest number of farmers), with almost a quarter (24.9 per cent) of farms doing so in 2016; up from 18.1 per cent in 2010 (the 2006 survey did not disaggregate 'rents from longer-term letting' to this level of detail). Providing tourist accommodation, renting out buildings for commercial use and providing agricultural services were the next most popular diversification activities (with between 10 and 15 per cent of respondents operating these) in all three years surveyed, followed by the provision of equine services and processing and/or retailing of farm produce.

Low returns from agriculture are also driving many farm households to seek income away from the farm. According to SW Farm Survey data from 2016, the average proportion of household income derived from agriculture (on the farms surveyed) was only 62.7 per cent. An average of 12.1 per cent came from non-farming enterprises on the farm, 12 per cent from pensions, savings and investments and 8.6 per cent from off-farm work. Only 21.5 per cent of respondents in 2016 said that all of their income came from agriculture on the farm (down from 22.7 per cent in 2010 and 30.9 per cent in 2006); and 17.3 per cent said that agriculture accounted for less than 25 per cent of their household income (up from 13.4 per cent in 2010 and 11.7 per cent in 2006). Together these findings point to an increasing disengagement from agriculture as a primary income source for significant numbers of South West farming households.

Changes to the farm business

As the aforementioned trends imply, farm businesses are not static but constantly undergoing change as farmers and farm households respond to financial, personal and regulatory imperatives. The 2016 SW Farm Survey included questions enabling closer analysis of change at the farm level and these revealed considerable changes being made to farm businesses in the preceding six years, particularly in relation to the amount of farm output and the number of livestock on the farm (Figure 3.11). Clearly, a considerable number of farms have been expanding, with almost a quarter (22.5 per cent) increasing the size of their farm and just over a third (35.6 per cent) increasing total farm output. Other farms, however, have curtailed their operations, with 12.2 per cent of respondents reducing the size of their farm and 23.3 per cent decreasing total farm output.

The greater numbers of farms reducing rather than increasing the amount of employed labour (16.5 per cent compared to 13.4 per cent) corresponds with the wider farm labour trends discussed earlier, and is largely mirrored by greater numbers of farms increasing than decreasing the amount of family labour (15.1 per cent compared to 11.4 per cent). There has also been a marked change in the amount of work contracted out; 19.7 per cent have increased this, compared to only 9.8 per cent decreasing it.

The 2006 and 2010 surveys did not ask respondents about changes they had already made. However, analysis of their planned changes for the succeeding five years indicates a similar pattern to that reported above in most areas, with

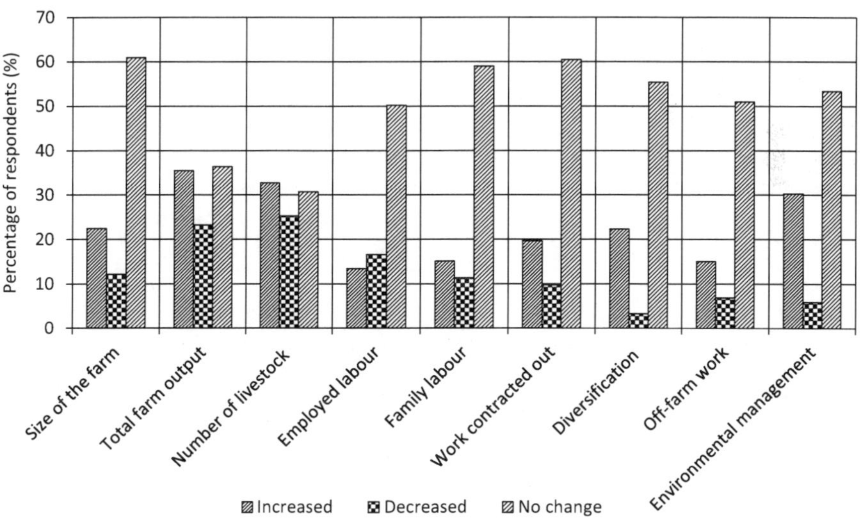

Figure 3.11 Farm business changes made since 2010

Source: SW Farm Survey

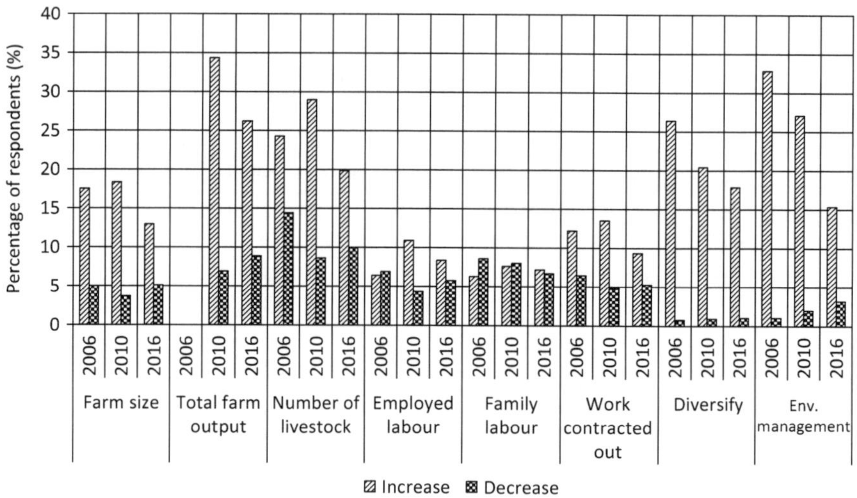

Figure 3.12 Farm business changes planned over the next five years in 2006, 2010 and 2016

Source: SW Farm Survey

overall net increases planned for farm size, output, livestock numbers, work contracted out and environmental management (Figure 3.12). Interestingly, however, the numbers of farmers planning increases to these aspects of their business were fewer in 2016 than in previous years, potentially indicating an overall weakening of confidence regarding the ability/willingness to expand and invest. Given that the 2016 survey was undertaken in the run-up to the UK's referendum on membership of the EU, it is entirely possible that it coincided with a period of low confidence associated with the unknown outcome of the vote and the difficulty in predicting the implications of a vote to leave for agricultural policy and the agricultural economy.

Also of note is the observation that in 2010 more farms planned to increase than reduce their amount of employed labour but there was actually a greater decrease in employed labour between 2010 and 2016, as reported by 2016 respondents. Although definitive conclusions cannot be drawn, since the 2010 and 2016 survey samples were not identical (though 812 respondents – 64.9 per cent of the 2016 sample – did complete both), this finding may indicate that optimism to employ more workers in 2010 was not able to be implemented in reality. Increased family labour may have filled planned labour requirements instead, as more farms increased than reduced family labour between 2010 and 2016 (in contrast to plans reported in 2010).

Although the survey data do not allow us to determine the causality of specific changes planned for the farm business, general factors motivating change were explored (Table 3.11). Unsurprisingly, in all three years surveyed,

Table 3.11 Factors influencing plans for change (% respondents influenced)

	2006	2010	2016
Single Farm Payment/Basic Payment Scheme	32.2	25.3	21.8
Farm profitability	54.1	55.5	52.5
Cost of inputs	33.8	32.5	27.3
Cost of borrowing	22.1	–	–
Availability of loans	–	12.0	8.1
Market prices	40.1	36.6	41.9
Environmental schemes	27.0	26.9	20.5
Family changes	21.7	24.8	28.4
Time of life	27.2	27.0	35.1
To make life easier	27.2	32.2	35.9
Food security	–	7.4	5.5
Climate change	–	6.5	8.8

Source: SW Farm Survey

'farm profitability' was the biggest influencer of change, with just over half of respondents citing this as a factor in all three surveys. Other explicitly financial factors – namely the cost of inputs and market prices – were also consistently notable influencers. However, social and behavioural factors are important; something that is often overlooked in accounts of farming change. For instance, in 2016, 'making life easier', 'time of life' and 'family changes' were all cited by around a third of respondents as factors influencing change. Notably, the influence of environmental schemes has declined since 2010 (cited as an influencer by 27 per cent and 26.9 per cent of respondents in 2006 and 2010 but only 20.5 per cent in 2016). This decline in influence is plausibly linked to the introduction of the new Countryside Stewardship Scheme in 2015, which suffered from poor uptake and, according to an NFU survey, was unpopular due to farmers perceiving it as bureaucratic, overly complex and providing insufficient payment for the actions involved (NFU, 2015a).

Leaving farming

Of respondents in the 2016 SW Farm Survey, 40.6 per cent said that they were planning to retire or leave farming in the next five years. In one sense this figure is unsurprising given that the median age of respondents was 60.9 and 46.4 per cent were aged 62 or over and thus reaching state pension age within the given timeframe. Indeed, 72.6 per cent of those saying they were planning on leaving farming in the next five years stated that 'getting too old' was one of the main reasons to do so. It is also notable that the mean age at which respondents were planning to retire (of those planning to do so in the next five years) was 69.7 in 2016; up from 66.7 in 2006 and 67.2 in 2010. These

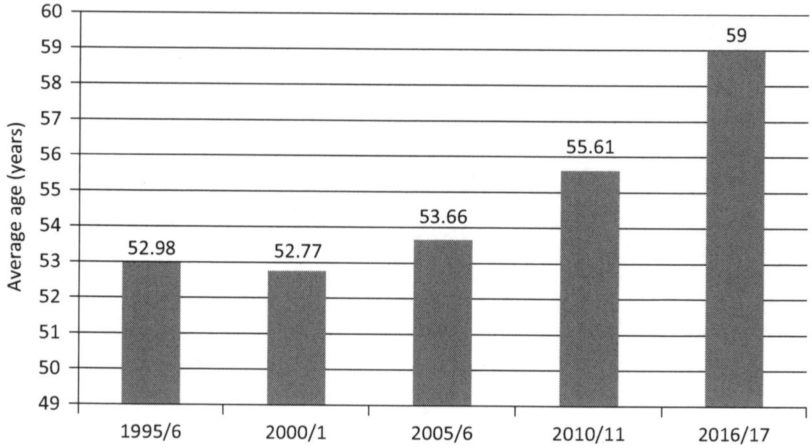

Figure 3.13 Average age of principal farmer in England, 1995–2017

Source: Defra & Rural Business Research, FBS 2016/17, 2018

data align with national figures that show the average age of principal farmers in England is increasing; see Figure 3.13.

The ageing of principal farmers and uncertainty about their ability to attract successors has led to concerns about the ability of family farming to reproduce itself in certain parts of the country (Chiswell and Lobley, 2018; Lobley et al., 2005; Lobley et al., 2012). While talk of a succession 'crisis' may be hyperbole (see Chiswell and Lobley, 2018), identifying and attracting a successor and then engaging in the process of succession planning with near inevitable 'difficult conversations' shifting patterns of power, responsibility and identity within the family, business and peer groups, and a typical disposition against long-term planning, means that successful succession cannot be taken for granted (Lobley et al., 2012).

While the ageing of principal farmers is not unique to the UK, neither is it a universal attribute of agriculture. A study of the 'young farmer problem' in Europe found that the proportion of young sole holders in the UK was particularly small compared to other countries given its relative abundance of large farms (since elsewhere young farmers were associated with larger farms) (Zagata and Sutherland, 2015). The authors suggest that this may be related to high land prices and a reluctance on the part of ageing farmers to transfer ownership in the UK, though they also note that 'the small number of young sole holders may mask a higher number of successors, who have yet to formally become the principle decision-makers on their farms' (Zagata and Sutherland, 2015, p.48). Nevertheless, an ageing population of farmers represents a concern to the overall productivity and sustainability of the industry, particularly as some research indicates that young farmers are associated with higher rates of innovation and technology adoption, tend to operate more efficient and economic farms, and are more likely to engage with agri-environment schemes (Diederen et al., 2003; Hamilton et al., 2015; Zagata and Sutherland, 2015).

Retirement, it has been argued, is a somewhat elusive concept for many farmers (Kirkpatrick, 2012). This can be for a multitude of reasons including: the important if somewhat mundane issue of a suitable income in retirement, which is often expected to derive in part at least from the farm (Whitehead et al., 2012); access to suitable housing (Winter and Lobley, 2016); and the oft-voiced reason that farming is a way of life rather than just a job. Often the incumbent will have spent much of their life, not just their career, learning about the farm, identifying with the farm, developing the farm, making sacrifices for the farm and taking risks for the farm, and the prospect of withdrawing from the leadership role can be both challenging and frightening as it may be associated with fears of 'the loss of power, status, or personal identity' (Sharma et al., 2003, p.671). The 'psychodynamic and sociodynamic factors that influence the process of transferring the family farm' (Conway et al., 2016, p.173) arguably provide a greater barrier to retirement than the more objective challenges of securing retirement income and the technical process of legally transferring ownership of assets.

Clearly age/retirement is not always the reason behind individuals' decisions to cease farming. Returning to our findings from the SW Farm Surveys, it is of concern that the percentage of farmers planning to leave farming in 2016 (40.6 per cent) is considerably greater than in 2006 (28.2 per cent) and 2010 (23.8 per cent) (though some of this increase is likely to be due to a higher average age of respondents; 60.9 in 2016 compared to 56.0 in 2006 and 57.3 in 2010). In 2016, of those planning to leave farming, 53.7 per cent of respondents said that they plan to retire from or leave farming in order to reduce physical work, 44.8 per cent wanted to have more time for other interests/holidays, and 34.9 per cent wanted to reduce stress. These responses attracted similarly high scores in 2006 and 2010 and, as we discuss in Chapter 8, are indicative of the impacts of stress and long hours of (often physically demanding) work on the health, well-being and quality of life of farmers.

A note on the limitations of official datasets

Most of the data presented so far in this chapter have been derived from official, large-scale surveys at the national level (i.e. the June Survey and FBS). These sources are useful in providing a broad picture of changes to the structure of the UK agricultural industry but the (lack of) detail they offer inevitably restricts the depth of analysis and, more importantly, can at times lead to misleading results. To some extent this is an inherent and unavoidable shortcoming of national-level data but, beyond this, we have also frequently found ourselves frustrated by limitations associated with underlying methodological variances and the extent of the surveys' coverage. We discuss some of these issues here not to discredit the use of these data, but to clarify its contribution and highlight the need to consider it alongside more detailed empirical evidence and analysis. Accordingly, the subsequent section draws on the academic literature in order to discuss in more detail issues of agricultural change and restructuring *at the farm level* – i.e. how national and regional trends translate into what is happening for the farmer on the ground.

One of the main difficulties associated with the June Survey is the way in which it represents farm size. It has long been the case that the June Census covered a large number of small, often non-commercial, farms, accounting for around 40 per cent of all farms prior to 2010 (Defra, 2011b). Defra has argued that the inclusion of these holdings presented an unrealistic picture of the agricultural industry. They 'resolved this issue by applying pre-defined thresholds to filter out these farms so the survey population now only covers the larger, more active farms', resulting in 'better indications of the genuine picture of structural change and the size of the active farming industry' (Defra, 2011b, p.4). As a result of the introduction of these pre-defined thresholds in 2010, non-commercial holdings have since been excluded from the survey and the census. Although Defra recognises this excludes a *significant number* of farm holdings, they estimate it only excludes around 1 per cent of main agricultural activities (e.g. sheep, crops, pigs) and improves the survey's ability to reveal genuine trends in agriculture. Nevertheless, we would argue that these aggregated data on farm size – which have long been the stock in trade for the analysis of structural change in agriculture – give an incomplete story and in some respects obscure the extent, pace and complexity of change. In particular:

- The data may underestimate the number of very small or micro holdings which fall below the June Census/Survey radar. This issue is likely to increase as a result of the 2015 decision to limit future CAP payments to holdings of 5 hectares or more. In that context there is little incentive for new micro holdings, which may arise from the dispersal of land where farms are split in farm sales, to register as a new holding.
- For tax reasons[10] many farmers appear in the June Survey as active businesses when in reality they are no longer actively farming. Others may let land for very short terms. The pressures on small to moderate sized farms, the need for economies of scale and fiscal rules have combined to encourage farmers to adopt a range of 'unconventional' occupancy arrangements, including 'grass keep', gentlemen's agreements, share farming and contract farming. In this respect the number of active small farms may be over-estimated and the number of larger farms underestimated. It is hard to estimate with precision the extent of these trends, but it is clear that the proportion of land held in unconventional arrangements has increased, accounting for 10.4 per cent of land in 1989 and 13.7 per cent in 2007 (Butler and Winter, 2008). Anecdotal evidence suggests that this trend has continued, probably

accelerated, and may well have been somewhat underestimated in Butler and Winter's research.

As Defra acknowledge, the introduction of thresholds has had a significant effect on estimates of farm labour, as a great many people are involved in agriculture on small, 'non-commercial' farms; for instance, in 2009, 19 per cent of the workforce were on the smallest holdings and many of these were part-time. Defra argue that 'in reality, as this work is only a hobby or spare-time activity, it is justified to exclude these from the estimates of the size of the agricultural labour force' (Defra, 2011b, p.5). This may be so, but it nevertheless paints an incomplete picture of the number of people involved in farming across the country.

In addition to the changes in farm size sampling, methodological changes to the June Survey in 2010 also included a change in the classification of farm types (to meet EU requirements) from Standard Gross Margins (SGM) to Standard Outputs (SO). As Defra acknowledges, this change prohibits longitudinal analysis of shifts in farm types pre- and post-2009/10. For instance, one of the most noticeable effects of the typology change is in the apparent substantive increase in the number of general cropping farms; this is, however, mainly due to 'specialist grass and forage' holdings (previously grouped under 'other' in the SGM typology) now being classified as general cropping (since grassland in these enterprises is considered to be 'forage for sale') under SO (Defra, 2011a). There are also noticeable differences in the apparent numbers of cereal and grazing livestock holdings as a result of this methodological change. Other changes to the methodology concerning farm types were also made in 2000, 2004 and 2006[11] (see Defra, 2011b), which, while having less of an impact on final results than the SGM to SO change, nevertheless further weaken the robustness of any longitudinal analysis relating to farm type.

The difficulty of piecing together a broad picture of structure and change in the UK agricultural industry is compounded by subtle but important differences in the underlying methodologies of the two key official data sources; the June Survey and FBS. For instance, in the June Survey,

> standard outputs are applied to the cropping and stocking as recorded on the survey day while in the FBS they are applied to the hectares of crop and average numbers of livestock over the year as a whole. Moreover, in the FBS, the minimum unit is a whole farm business, which may comprise more than one holding, while in the June Survey the holdings making up a farm may be treated separately.
>
> (Defra, 2017, p.26)

Although such differences do not preclude the use of the two datasets alongside each other, they do add to the long list of caveats regarding wider interpretation and comparison of the data.

These issues aside, perhaps a more fundamental concern regarding the reliance on these official datasets is that their focus on the physical and economic characteristics of the agricultural industry marginalises, or at least detracts attention from, the social and cultural aspects of farming and the people who lie at the heart of the matter – i.e. farmers and farming families. Trends in farm sizes, types, labour and income say nothing about what is going on at the farm level and what the implications are for day-to-day experiences of farming. The discussion below thus begins to consider the socio-economic and 'human' aspects of agricultural restructuring; a task which we also take up further in our focus on social change and well-being in Chapter 7.

Restructuring, adaptation and resistance to change at the farm level

Having considered the evidence on patterns of agricultural change from official data sources, we now turn to explore insights from the academic literature. A growing body of research in the UK and across western Europe (Calleja et al., 2012; de Roest et al., 2017; Dervillé et al., 2017; Evans, 2009; Lobley and Potter, 2004; Marsden et al., 2002; Richards et al., 2013; Zimmermann and Heckelei, 2012) suggests that the pattern of contemporary agricultural change is both complex and geographically differentiated. A long and sometimes difficult process of change has been taking place that has seen, for some, disengagement from agriculture as a mainstream income source, while others have specialised and expanded their farm business, and still others have struggled to adapt to changing economic conditions, changing policy signals and the changing position of farmers within society. The result of this complex pattern of adaptation and resistance to change is an agricultural land management community that is increasingly diverse in its social composition and behaviour (Lobley and Potter, 2004). These processes and the associated outcomes and implications are spatially uneven. The concept of spatial differentiation in agriculture and farming change is well established (Lange et al., 2013; Lobley and Potter, 2004; Marsden, 1998; Wilson, 2008) and it has been argued that the period between the 1980s and 1990s saw a greater regional polarisation of farming activities with the east of the country become more arable and the arable of pasture expanding in the south-west and north[12] (Haines-Young and McNally, 2001). The differentiated countryside exists not just in terms of farming patterns but also in terms of socio-economic characteristics, farmer and farm household behaviour and also both adaptability and vulnerability to future change. The opportunities and constraints faced by farmers and land managers are spatially variable and the resulting divergent patterns of restructuring are contributing to an increasingly differentiated countryside. In other words, in understanding agricultural change, place matters.

The conceptualisation in policy studies of agricultural restructuring as an externally 'driven' process (Entec, 2000; Haines-Young and McNally, 2001) implies significant, even radical, change – both in the past and yet to come. At the turn of the century, Savills predicted that, 'change in agriculture will accelerate further from its existing accelerating trajectory, suggesting exponential increases in restructuring if farm income does not improve' (Savills, 2001, p.5). To a certain extent, as the data we present in this chapter attest, this forecast has been borne out in the intervening years and is likely to remain pertinent in the context of post-Brexit changes to UK agricultural policy and markets. However, the limited numbers of empirical studies of farm household adjustment tell a more complex story than these macro-scale statistics are able to portray, with longstanding and persisting evidence both of adaptation and resistance to change, as well as considerable heterogeneity in the adoption of more radical restructuring approaches.

Gasson et al. (1998) for instance, explored the steps taken in response to the financial uncertainties of the 1990s. In common with the results of earlier surveys (e.g. Errington and Tranter, 1991), the responses recorded illustrate a strong commitment to production and specialisation. The most common response for many farm households being to follow a tried and tested strategy of increasing output from existing enterprises, cutting out unprofitable enterprises while reducing input, machinery and labour costs. Despite the wide range of changes implemented, many were resistant to adjusting the development trajectory of their business with almost half claiming to be essentially carrying on as before. Evidence from a case study survey of farmers in north-west Devon (Reed et al., 2002) also identified a clear resistance to change, among livestock farmers in particular, with many surviving only through a combination of 'belt-tightening', avoidance of risk and consumption of savings.

Further evidence of resistance to change – as well as a certain resilience to adversity – emerges in the responses of farmers to external shocks such as extreme weather and animal disease, which can have a huge economic impact on agriculture. For instance, widespread floods in the UK in the winter of 2014 were estimated to have a total economic cost to agriculture of £20.9 million,[13] or £470 per hectare flooded (ADAS, 2014). The impact of animal disease can likewise be crippling. The major outbreak of foot and mouth disease (FMD) in 2001, for example, was estimated to have cost the private sector a total of £5 billion (National Audit Office, 2002). Notably, though, the costs of such crises are widely variable from farm to farm, depending on a range of factors relating to the farm's structure and business model. For instance, in an analysis of the impact of bovine tuberculosis (bTB) on case study farms in the South West, Butler et al. (2010) found that the monthly loss of a bTB breakdown varied from just under £505 to nearly £3,184.

Despite such setbacks, however, many farms strive to continue their enterprises in a similar manner as to before. Research undertaken in Wales during the FMD outbreak in August 2001 (Christie et al., 2002) pointed to a relatively small increase in those planning to leave farming (mostly a result of bringing retirement decisions forward) and a general willingness to restock, although not necessarily to pre-FMD levels. Indeed, evidence from the FBS suggests that between 30 and 60 per cent of culled animals had been re-stocked by the end of the 2001/2 accounting year (Defra, 2002). Research in Cumbria by Barnett et al. (2002) produced similar findings, with 63 out of 67 farmers saying they would continue farming after the outbreak and only one saying they would definitely cease. Others, on the other hand, do use such opportunities (either willingly or unwillingly) to modify their operations: a quarter of Barnett et al.'s sample said that they would possibly or definitely expand their operations, though another quarter said that they would possibly or definitely reduce their farming activity (and were firmer in their intentions to do so). Evidence from a 2001 survey for Defra by ADAS (conducted in England) also suggested an increase in those planning to leave farming as well as an increase in those planning to adapt through diversification (Francis et al., 2002).

Research suggests that in order to understand different patterns of adaptive response, it is important to recognise more explicitly farmers' differing dispositions to act. Shucksmith and Herrmann (2002), for instance, explored differences between farm households based on a series of variables relating to farm and farm household characteristics, farmer attitudes and farmer socialisation. The resulting six distinct types of farming and farm household situation reflect varying degrees of resistance to change and willingness to adapt with implications for the future vulnerability of some of the groups identified. Some, such as the 'hobby farmers' and 'pluriactive successors' were largely disengaged from agriculture as a primary income source and were unresponsive to any other than the most radical policy signals. The 'struggling' and 'contented monoactives' on the other hand, reflected the findings of Gasson et al. (1998, p.44) reported above in that they 'favoured tried and tested methods of farming rather than innovation', were very conformist in terms of investment decisions and were resistant to change. Finally, two other groups had a distinctive approach to their farm, or more precisely, their farm resources. 'Potential diversifiers' and 'agribusinessmen' differed in some respects but both groups were characterised by a flexible approach to business development and viewed their farms 'as a collection of resources that could be deployed and redeployed in search of maximum profit' (ibid., p.46). This type of 'portfolio entrepreneurship' (Carter, 2001) can play an important role in stimulating rural employment and enterprise but also suggests the potential for more rapid restructuring and a shift of resources out of agriculture.

More recently, Morris et al. (2017) have also emphasised the heterogeneity of farm approaches to restructuring, arguing that the particular pathway chosen is influenced by a number of factors including available resources, geographical location, access to technology and the skills of the farmer, as well as considerations relating to identity, social context and the circumstances of the farming household. Based on qualitative interviews and quantitative questionnaire data in Wales, the authors characterise diversification strategy according to four farmer types. *High resource maximisers* seek a range of income streams from on and off the farm, utilise all available resources, embrace new technologies, and demonstrate personal entrepreneurial attributes. *Farm-focused* individuals also embrace technologies and demonstrate personal entrepreneurial attributes, but focus on the farm business and maximising farm resources rather than on alternative income streams. In contrast to these groups, *passive farmers* display low levels of engagement with technology, limited engagement with grant-focused opportunities and persist with mainstream farming systems/methods. *Lifestyle farmers* also demonstrate low levels of technology adoption but are supported by off-farm income and focus on promoting their lifestyle opportunities by pursuing a farming style adapted to their lifestyle/value choices. In general, entrepreneurial farmers (i.e. high resource maximisers and farm-focused individuals) were associated with larger farms, higher levels of education and a younger age profile than passive and lifestyle farmers. This research confirms the continued and growing importance of non-traditional sources of income to the economic livelihood of many farm households, while highlighting that the capacity and/or inclination to optimise such income is dependent on a number of internal and external factors. Farmers are revealed here as 'reactive dynamic business operators, not only facing the physical constraints of weather and land, but also influenced and constrained by social and policy expectations' (Morris et al., 2017, p.141).

In sum, while there is evidence that many family farms simply attempt to 'absorb' change by making internal, farm household adjustments (Lobley and Potter, 2004; Reed et al., 2002), others are taking more active steps to respond by making decisions about the deployment of resources either singularly or in combination. In turn, these restructuring decisions have a range of implications for the rural environment, economy and society. Studies by Burton et al. (2005) and by Appleby (2004), for instance, have pointed to the decline of 'social capital' in UK farming due to an erosion of community ties and collective working arrangements. Although a study by Williams (2002) on changing patterns of community participation does not focus on farmers as such, it suggests some potentially significant linkages between agricultural restructuring and the willingness and ability of farmers to contribute to community life.

Conclusion

Clearly, a wide range of forces are currently driving agricultural restructuring, ranging from the process of globalisation and the marked economic downturn which has characterised much of British agriculture in recent years, to changes during the life course at the farm household level and unanticipated events such as animal disease outbreaks and extreme weather. To this list we can now add the uncertainty around Brexit (at the time of writing) and in coming years the impact of a new British Agricultural Policy and new international trading conditions.

Data from official sources such as the annual June Survey and FBS, while important in charting the broad contours of agricultural change, fail to capture the complexity of agricultural change and restructuring and tell us little about the overall development pathways of family farms. The data presented in this chapter on diversification and off-farm work reflect findings from the wider literature that confirm increasing pluri-activity and significant heterogeneity in farm business development pathways as farms seek to broaden their income base in different ways (Evans, 2009; McFadden and Gorman, 2016; Morris et al., 2017; Wilson, 2008). Such restructuring has, in part, also been influenced by the decoupling of CAP support payments since 2003 and other policy drivers urging farmers to meet social and environmental, as well as production, objectives (Morris et al., 2017). Naturally, though, precise restructuring responses and levels of diversification and innovation are dependent on the specificities of the farm business and the personalities and identities of the individuals involved (McFadden and Gorman, 2016; Shucksmith and Herrmann, 2002). In some areas, competition for land from 'residential' and 'lifestyle' farmers may well force existing 'mainstream' farm operators to seek alternatives to the more traditional routes of restructuring and survival through expansion, while in other areas the development of large-scale contract farming businesses may reflect the new face of farming expansion. On the other hand, previous research suggests that many farmers stick to tried and tested practices, or simply attempt to 'absorb' change by making internal, farm household adjustments (Reed et al., 2002), with the decision to leave farming taken only after all other alternatives have been exhausted (Errington and Tranter, 1991). Whether through restructuring, adaptation or resistance to change, what is clear is that a number of different strategies are being followed.

The adjustment to farming practice, living standards and lifestyles which all this implies is not without personal cost. While media claims of an agricultural crisis may be exaggerated, it is clear that large numbers of farmers are finding they have to make difficult adjustments against a shifting background of policy reform and market change. Even successful transformations can have negative implications for the lives of the farmer and his/her family. For instance, a reduction in the number of hired workers on farms means that farmers often find themselves working alone, or with members of their immediate family, rather than as part of a team. Meanwhile, the requirement for farmers' spouses to go out to work to generate additional household income leaves the farmer in isolation for large parts of the working day. We return to discuss these issues more fully in Chapter 8.

The implications for UK agriculture of the decision to leave the EU are as yet unclear, but the nature of the nation's relationship with the rest of Europe and the shape of its agricultural policy in a post-Brexit era will have a powerful influence on the types of trends and patterns discussed in this chapter. The outcomes of negotiations regarding trade arrangements, for instance, could have a significant impact on how land is managed, as any changes to tariffs and regulations will influence the production outputs and methods chosen by farmers (Franks, 2016). Furthermore, while trends towards market liberalisation and subsidies that support environmental and rural development objectives are expected to continue post-Brexit (Whitfield and Marshall, 2017), inevitable changes to the design and process around such funding and the ultimate size of the agricultural budget are bound to have an impact on farm businesses, households and the management decisions they take. We return to the issue of Brexit in the final chapter. In the next chapter we discuss the all-important role of the market, ranging from the revolution in food retailing through to the (re)emergence of direct sales.

Notes

1 The June Survey of Agriculture and Horticulture has a long lineage in England. First run in 1866, it was carried out as a full annual census until 1995 when it was reduced to a sample survey, surveying around 80 per cent of the farming population. The survey now samples between 30,000 and 70,000 holdings each year and is designed to collect detailed information on agricultural activities and the agricultural labour force. A full census is carried out once every 10 years, 2010 being the most recent. From 2011 the survey has been run predominantly online.

2 Standard Gross Margin (SGM) is a measure of profitability, whereas Standard Output (SO) is a measure of the total value of the output costs on a farm based on standardised coefficients.

3 Figures prior to 2009 relate to all holdings, whereas figures from 2009 onwards relate to commercial holdings only. The 2009 figures were revised in 2010 to be comparable with future years.

4 FBI includes returns from agriculture, agri-environment, diversification and the Single Farm Payment. It does not include the value of own labour or the rental value of owned land.

5 Return on capital employed (ROCE) is a measure of the return that a business makes from the available capital. ROCE provides a more holistic view than profit margins, focusing on efficient use of capital and low costs and allowing an equal comparison across farms of differing sizes. A positive ROCE value shows that a farm is achieving an economic return on the capital used. ROCE = Earnings before Interest and Tax ÷ Total Assets less Current Liabilities (Defra, 2016, p.19).

6 FBS farm size categories are based on Standard Labour Requirements (SLR), as follows: Small = 1 to < 2 SLR; Medium = 2 to < 3 SLR; Large = 3 < 5 SLR; Very large = 5+ SLR.

7 The authors note that some of the variation in FBI is due to FBI being calculated before unpaid labour and notional rent under the Farm Business Survey (which will have a disproportionate reducing effect on large farms that rely less on unpaid family labour), but that this cannot account for the extent of the variation in FBI.

8 Defra explain total factor productivity as; 'a measure of how well inputs are converted into outputs giving an indication of the efficiency and competitiveness of the agriculture industry' (Defra and Office for National Statistics, 2017, p.1).

9 Average annual growth in the UK between 2002 and 2012 was 0.8 per cent compared to 2 per cent in the US, 1.8 per cent in Germany and 1.7 per cent in France.

10 There are tax advantages to land owners who no longer wish to actively farm to use share farming, contract farming, partnerships, seasonal leases and licences to demonstrate continuing trading activity when in practice they may take no risk and lack any management control.

11 Prior to 2000, results were based on main holdings only. In 2004, changes were made to the farm type (e.g. from 'cattle and sheep' to 'grazing livestock', as explained under Table 3.1) and there was a switch from using 1988 to 2000 SGM coefficients. From 2006 onwards cattle data is derived from the Cattle Tracing System rather than the survey data (Defra, 2011b).

12 Since then the pattern of ever greater polarisation appears to have halted and although the broad scale difference between the regions has been largely maintained, June Survey data (Defra, 2018b) indicates that the grassland area in the East has increased since in mid-1990s, whereas the South West has seen a decline in grass area and an increase in arable land (mostly likely for home-grown livestock feed).

13 £20.9 million is the central estimate. The lower range estimate was £13.2 million and the higher range £28.5 million.

References

ADAS (2014). *The Economic Impact of the 2014 Winter Floods on Agriculture in England.* London: Defra.

AHDB (2017a). *Average herd size.* [Online] Available at: https://dairy.ahdb.org.uk/resources-library/market-information/farming-data/average-herd-size/#.WcpePLKGOUk [accessed 10/10/17].

AHDB (2017b). *Average milk yield.* [Online] Available at: https://dairy.ahdb.org.uk/resources-library/market-information/farming-data/average-milk-yield/#.Wcpiw02WyUk [accessed 10/10/17].

AHDB (2017c). *Cow numbers.* [Online] Available at: https://dairy.ahdb.org.uk/resources-library/market-information/farming-data/cow-numbers/#.WcpkWbKGOUk [accessed 10/10/17].

AHDB (2017d). *UK farm to retail price spreads.* [Online] Available at: http://beefandlamb.ahdb.org.uk/markets/industry-reports/uk-statistics/ [accessed 10/10/17].

Appleby, M. (2004). *Norfolk Arable Land Management Initiative (NALMI). Final project report.* London: Countryside Agency.

Barnett, K., Caroll, T., Lowe, P. & Phillipson, J. (2002). *Coping with Crisis in Cumbria: The consequences of foot and mouth disease.* University of Newcastle: Centre for Rural Economy.

Bartolini, F. & Viaggi, D. (2013). The Common Agricultural Policy and the determinants of changes in EU farm size. *Land Use Policy*, 31, 126–135.

Bowler, I. (2002). Developing sustainable agriculture. *Geography*, 87(3), 205–212.

Brassley, P. (1998). On the unrecognized significance of the ephemeral landscape. *Landscape Research*, 23(2), 119–132.

Burton, R., Mansfield, L., Schwartz, G., Brown, K. & Convery, L. (2005). *Social Capital in Hill Farming. Report for the Upland Centre*. Aberdeen: Macaulay Institute.

Butler, A., Lobley, M. & Winter, D. M. (2010). *Economic Impact Assessment of BovineTuberculosis in the South West of England*. CRPR Research Paper No.30. University of Exeter: Centre for Rural Policy Research.

Butler, A. & Winter, M. (2008). *Agricultural Tenure in England and Wales 2007*. University of Exeter: Centre for Rural Policy Research.

Butler, S. J., Vickery, J. A. & Norris, K. (2007). Farmland biodiversity and the footprint of agriculture. *Science*, 315(5810), 381–384.

Calleja, E. J., Ilbery, B. & Mills, P. R. (2012). Agricultural change and the rise of the British strawberry industry, 1920–2009. *Journal of Rural Studies*, 28(4), 603–611.

Carter, S. (2001). Multiple business ownership in the farm sector – Differentiating monoactive, diversified and portfolio enterprises. *International Journal of Entrepreneurial Behavior & Research*, 7(2), 43–59.

Chiswell, H. M. & Lobley, M. (2018). 'It's definitely a good time to be a farmer': Understanding the changing dynamics of successor creation in late modern society. *Rural Sociology*, Early View.

Christie, M., Scott, A., Midmore, P. & Youell, R. (2002). *The Impacts of Foot and Mouth Disease on Welsh Farming. Report to the Countryside Council for Wales*. University of Wales: Institute of Rural Studies and the School of Management and Business.

Conway, S. F., McDonagh, J., Farrell, M. & Kinsella, A. (2016). Cease agricultural activity forever? Underestimating the importance of symbolic capital. *Journal of Rural Studies*, 44, 164–176.

de Roest, K., Ferrari, P. & Knickel, K. (2017). Specialisation and economies of scale or diversification and economies of scope? Assessing different agricultural development pathways. *Journal of Rural Studies*, 59, 242–251.

Defra (2002). *Impact of Foot and Mouth Disease on UK Agriculture (in Farm Incomes 2001/2)*. London: Defra.

Defra (2008). *Agricultural Specialisation. Agricultural Change and Environment Observatory Research Report No.11*. London: Defra.

Defra (2011a). *Farming Statistics Note on the Revised EC Classification of Farm Types: Effects on the June Survey population and Farm Business Survey sample in England*. London: Defra.

Defra (2011b). *June Survey of Agriculture and Horticulture: Methodology*. London: Defra.

Defra (2013). *Farm Practices Survey Autumn 2012 – England*. London: Defra.

Defra (2014). *Contracting on English Farms: Evidence from existing surveys. Agricultural Change and Environment Observatory Research Report No. 35*. London: Defra.

Defra (2016). *Balance sheet analysis and farming performance, England 2014/15*. London: Defra.

Defra (2017). *Farm Accounts in England – Results from the Farm Business Survey 2016/17*. London: Defra.

Defra (2018a). *Farm Business Survey*. [Online] Available at: www.gov.uk/government/collections/farm-business-survey [accessed 12/03/18].

Defra (2018b). *Structure of the agricultural industry in England and the UK at June*. [Online] Available at: www.gov.uk/government/statistical-data-sets/structure-of-the-agricultural-industry-in-england-and-the-uk-at-june [accessed 12/03/18].

Defra, Department of Agriculture Environment and Rural Affairs (Northern Ireland), Welsh Assembly Government & The Scottish Government (2017). *Agriculture in the United Kingdom 2016*. London: Defra.

Defra & Government Statistical Service (2018). *The Future Farming and Environment Evidence Compendium*. London: Defra.

Defra & Office for National Statistics (2017). *Total Factor Productivity of the UK Agriculture Industry. 2016, 1st estimate – Statistical notice*. London: Defra.

Defra & Rural Business Research (2018). *Farm Business Survey (FBS) data builder*. [Online] Available at: www.farmbusinesssurvey.co.uk/DataBuilder/ [accessed 05/02/18].

Dervillé, M., Allaire, G., Maigné, É. & Cahuzac, É. (2017). Internal and contextual drivers of dairy restructuring: Evidence from French mountainous areas and post-quota prospects. *Agricultural Economics*, 48(1), 91–103.

Diederen, P., Van Meijl, H., Wolters, A. & Bijak, K. (2003). Innovation adoption in agriculture: Innovators, early adopters and laggards. *Cahiers d'Economie et de Sociologie Rurales*, 67, 29–50.

Downing, E. (2016). *UK Dairy Industry – Current Issues and Challenges*. London: House of Commons Library.

Entec UK Ltd. (2000). *Drivers of Change and Future Scenarios*. Cheltenham: Countryside Agency.

Errington, A. & Tranter, R. (1991). *Getting Out of Farming? Part two: The Farmers*. University of Reading: Farm Management Unit.

Evans, N. (2009). Adjustment strategies revisited: Agricultural change in the Welsh Marches. *Journal of Rural Studies*, 25(2), 217–230.

Francis, J., Hawley, J. & Scott, T. (2002). *Report on a Survey of the Effects of the 2001 Foot & Mouth Disease outbreak (England)*. London: Defra.

Franks, J. (2016). *Some Implications of Brexit for UK Agricultural Environmental Policy*. Centre for Rural Economy Discussion Paper Series No. 36. Newcastle University: Centre for Rural Economy.

Gasson, R., Errington, A. & Tranter, R. (1998). *Carry On Farming: A study of how English farmers have adapted to the changing pressures on farming*. London: Imperial College Press.

Goddard, E., Weersink, A., Chen, K. & Turvey, C. G. (1993). Economics of Structural Change in Agriculture. *Canadian Journal of Agricultural Economics/Revue canadienne d'agroeconomie*, 41(4), 475–489.

Haines-Young, R. & McNally, S. (2001). *Drivers of Countryside Change*. Huntingdon: Centre for Ecology & Hydrology.

Hamilton, W., Bosworth, G. & Ruto, E. (2015). Entrepreneurial younger farmers and the "young farmer problem" in England. *Agriculture and Forestry*, 61(4), 61–69.

Harvey, D. R. (1989). The economics of the farmland market. Paper presented to *Agricultural Economics Society One Day Conference: The agricultural land market*. The Royal Society, London, 15 December, 1989.

Hoggart, K. & Paniagua, A. (2001). What rural restructuring? *Journal of Rural Studies*, 17(1), 41–62.

Ilbery, B., Ingram, J., Kirwan, J., Maye, D. & Prince, N. (2009). *Structural Change and New Entrants in UK Agriculture: Examining the role of county farms and the fresh start initiative in Cornwall.* Gloucester: Countryside and Communities Research Institute.

Ilbery, B. & Maye, D. (2010). Agricultural restructuring and changing food networks in the UK. *In:* Coe, N. & Jones, A. (eds) *The Economic Geography of the UK.* London: SAGE.

Kirkpatrick, J. (2012). Retired farmer – An elusive concept. *In:* Lobley, M., Baker, J. & Whitehead, I. (eds) *Keeping it in the Family: International Perspectives on Succession and Retirement on Family Farms.* Abingdon: Ashgate, 165–178.

Lange, A., Piorr, A., Siebert, R. & Zasada, I. (2013). Spatial differentiation of farm diversification: How rural attractiveness and vicinity to cities determine farm households' response to the CAP. *Land Use Policy*, 31, 136–144.

Lobley, M., Baker, J. & Whitehead, I. (2012). *Keeping it in the Family: International perspectives on succession and retirement on family farms.* Abingdon: Ashgate.

Lobley, M. & Potter, C. (2004). Agricultural change and restructuring: Recent evidence from a survey of agricultural households in England. *Journal of Rural Studies*, 20(4), 499–510.

Lobley, M., Potter, C., Butler, A., Whitehead, I. & Millard, N. (2005). *The Wider Social Impacts of Changes in the Structure of Agricultural Businesses.* University of Exeter: Centre for Rural Policy Research.

Marsden, T. (1998). New rural territories: Regulating the differentiated rural spaces. *Journal of Rural Studies*, 14(1), 107–117.

Marsden, T., Banks, J. & Bristow, G. (2002). The social management of rural nature: Understanding agrarian-based rural development. *Environment and Planning A*, 34(5), 809–825.

McFadden, T. & Gorman, M. (2016). Exploring the concept of farm household innovation capacity in relation to farm diversification in policy context. *Journal of Rural Studies*, 46, 60–70.

Morris, W., Henley, A. & Dowell, D. (2017). Farm diversification, entrepreneurship and technology adoption: Analysis of upland farmers in Wales. *Journal of Rural Studies*, 53, 132–143.

MSCI (2015). *IPD UK Annual Rural Property Index*. New York: MSCI.

Murdoch, J. & Marsden, T. (1994). *Reconstituting Rurality: Class, community and power in the development process.* London: UCL Press.

National Audit Office (2002). *The 2001 Outbreak of Foot and Mouth Disease*. London: NAO.

NFU (2015a). *CS scheme 'too complex' – survey reveals* [Online]. Available at: www.nfuonline.com/news/latest-news/cs-scheme-too-complex-survey-reveals/ [accessed 11/10/17].

NFU (2015b). *UK agricultural productivity fails to keep pace with global trends* [Online]. Available at: www.nfuonline.com/cross-sector/farm-business/economic-intelligence/economic-intelligence-news/uk-agricultural-productivity-fails-to-keep-pace-with-global-trends/ [accessed 16/10/17].

Nye, C. (2018). The 'blind spot' of agricultural research: Labour flexibility, composition and worker availability in the South West of England. *Cahiers Agricultures*, 27(3), 35002.

Nye, C., Lobley, M. & Winter, D. M. (2017). *LEEP – University of Exeter – Written Evidence in Response to the UK Parliamentary Inquiry, Feeding the Nation: Labour constraints* London: Houses of Parliament.

Oltmans, A. W. (2007). A new approach in farm business analysis to fit a changing farmland investment market. Paper presented to *16th International Farm Management Association Congress.* Cork, Ireland, 15–20 July 2007.

Rackham, O. (1986). *The History of the Countryside.* London: Phoenix Press.

Reed, M., Lobley, M., Winter, M. & Chandler, J. (2002). *Family Farmers on the Edge: Adaptability and change in farm households.* University of Plymouth: Department of Land Use and Rural Management.

Richards, C., Bjørkhaug, H., Lawrence, G. & Hickman, E. (2013). Retailer-driven agricultural restructuring—Australia, the UK and Norway in comparison. *Agriculture and Human Values*, 30(2), 235–245.

RICS Economics (2017). *RICS/RAU Rural Land Market Survey H1 2017: Lack of clarity over post-Brexit landscape still a concern in the market.* London: RICS Economics

Savills (2001). *Structural Change and the Implications for the Countryside. Report to the UK countryside agencies.* Peterborough: Savills.

Savills (2013). *Market Survey: UK agricultural land.* London: Savills.

Savills (2015). *Market Survey: UK agricultural land.* London: Savills.

Savills (2016). *Arable Benchmarking Survey: Harvest 2015.* London: Savills.

Savills (2017a). *GB Farmland Market.* London: Savills.

Savills (2017b). *Market Survey: GB agricultural land.* London: Savills.

Savills (2018). *Spotlight 2018: GB agricultural land.* London: Savills.

Savills & Duchy and Bicton College's Rural Business School (2017). *Where Dairy Farm Cash Flows.* London: Savills.

Sharma, P., Chrisman, J. J. & Chua, J. H. (2003). Predictors of satisfaction with the succession process in family firms. *Journal of Business Venturing*, 18(5), 667–687.

Shucksmith, M. & Herrmann, V. (2002). Future changes in British agriculture: Projecting divergent farm household behaviour. *Journal of Agricultural Economics*, 53(1), 37–50.

Stoate, C., Báldi, A., Beja, P., Boatman, N. D., Herzon, I., van Doorn, A., de Snoo, G. R., Rakosy, L. & Ramwell, C. (2009). Ecological impacts of early 21st century agricultural change in Europe – A review. *Journal of Environmental Management*, 91(1), 22–46.

University of Leicester. (2015). *State of our countryside: Land use map of the United Kingdom reveals large-scale changes in environment* [Online]. Available at: www2.le.ac.uk/offices/press/press-releases/2015/june/state-of-our-countryside-land-use-map-of-the-united-kingdom-reveals-large-scale-changes-in-environment [accessed 10/10/17].

Whitehead, I., Lobley, M. & Baker, J. (2012). From generation to generation: Drawing the threads together. *In:* Lobley, M., Baker, J. & Whitehead, I. (eds) *Keeping it in the Family: International perspectives on succession and retirement on family farms.* Oxford: Ashgate, 213–40.

Whitfield, S. & Marshall, A. (2017). Defining and delivering 'sustainable' agriculture in the UK after Brexit: Interdisciplinary lessons from experiences of agricultural reform. *International Journal of Agricultural Sustainability*, 15(5), 501–513.

Williams, C. C. (2002). Harnessing community self-help: Some lessons from rural England. *Local Economy*, 17(2), 136–146.

Wilson, G. A. (2008). From 'weak' to 'strong' multifunctionality: Conceptualising farm-level multifunctional transitional pathways. *Journal of Rural Studies*, 24(3), 367–383.

Winter, M. & Lobley, M. (2016). *Is there a Future for the Small Family Farm in the UK?* Report to the Prince's Countryside Fund. London: Prince's Countryside Fund.

Woods, M. (2005). *Rural Geography: Processes, responses and experiences in rural restructuring.* London: SAGE.

Zagata, L. & Sutherland, L.-A. (2015). Deconstructing the 'young farmer problem in Europe': Towards a research agenda. *Journal of Rural Studies*, 38, 39–51.

Zimmermann, A. & Heckelei, T. (2012). Structural change of European dairy farms: A cross-regional analysis. *Journal of Agricultural Economics*, 63(3), 576–603.

4 Farmers and the market

Introduction

A sustainable future for farmers is only possible if they can sell their products, and do so at prices that will ensure both adequate returns on investment in terms of farm household livelihoods and reinvestment in the farm business. Long gone are the days when direct marketing to consumers provided for the majority of such returns in a UK context. Therefore the production and manufacture of food products require outlets and systems for the sale of food. There are a huge variety of methods for the retailing of food and food products and these vary considerably over space and time. In this chapter we examine the retail revolution and the emergence of the giant retail companies, which are particularly powerful in the UK. Where, when and how we shop for food and the products on offer to us is now shaped by a small number of large multiple retail companies, whose outlets are often sited in purpose-built superstores on the edge of town. In the UK, the food retailing giants such as Sainsbury's, Tesco and Morrisons are inescapably central to the lives of the majority of modern food consumers. These retailers have not only reshaped the food shopping experience of the individual customer and radically altered the built environment of suburban landscapes but have also influenced profoundly the market opportunities and experiences of farmers. The chapter, therefore, commences with an overview of the retail revolution. It turns then to the implications of these changes for farmers.

We also examine the emergence of alternatives as the retailers face criticism for their style and the extent of their market share. Of course, there are examples where direct marketing may continue as an adjunct to commercial farming systems. The old pannier markets of Britain may have largely died out, but many livestock auction markets still retain a section where small quantities of local vegetables, eggs and poultry for the table may be sold by auction to local residents, although this is increasingly subject to restrictions on public health and hygiene grounds. Throughout Europe, there are farmers willing to sell small quantities of produce direct to local consumers at the door. Fruit and vegetables are the most commonly sold items in this way. The direct sale of dairy and meat products now faces formidable difficulties from hygiene regulations in most places, which mean, in essence, that prior to sale the farmer has to engage in various stages of processing. The extent to which farmers are seeking to return to direct sale is returned to later in the chapter.

The retail revolution

The shift from sole-trader outlets for food retail in the UK started in the mid-1900s with the slow growth of retail chains. The economic benefits of multiple food retailing ensured that this business model had a large share of the mid-twentieth-century retail landscape. In Britain between the wars, companies with multiple grocery shops grew rapidly, due in part to cheap, imported foods. It is estimated that in 1939 co-operative retail societies accounted for as much as 24 per cent of the total sales in the retail groceries and provisions market and multiple retailers for as much as 25 per cent (Jeffreys, 1954). Important though these developments were, the co-ops and multiples continued to operate in behind-the-counter formats very similar to those of the independents. In the UK, self-service shopping took longer to take hold than in north America where it began in the 1900s. As late as 1947, there were just ten self-service shops in the country (Shaw et al., 2004; quoting Fulop, 1964, p.19), but by 1959 Shaw et al. (2004) estimated there were 5,850. In early post-war Britain, food rationing, continuing regulation of building materials and Resale Price Maintenance (RPM) all contributed to restricting the development of self-service and, more especially, the adoption of American-style supermarkets (Shaw et al., 2004; Shaw et al., 2012). The UK's first supermarket under the new Premier Supermarkets brand, which opened in South London in 1951, soon grossed ten times as much per week as the average British general store of the time (Gregory, 2010; Stanton, 2018). Prior to this retail revolution, grocery shopping entailed visiting several different shops located near to one another in the high streets of towns and cities, each selling a particular food product e.g. bread, fish, fruit and vegetables, general groceries and meat. Customers would be served by a shop keeper from behind a counter. The emergence of the self-service and/or supermarket, where the whole range of food products was made available under one roof, undermined this pattern of retailing. The development was novel for both customer and retailer. For customers, instead of asking for items over the counter they collected the goods themselves and paid at a check-out. This opened up new possibilities of inspecting goods (or rather their packaging!) in more detail prior to purchase, and increased the possibility of trying out new goods. For the retailer it represented a revolution in selling technique allowing both for the cutting of labour costs and the likelihood of increased sales. The appearance of the self-service shop or supermarket placed the high street retailers under pressure. The retail revolution had begun (Mathias, 1967).

The contribution of food retailing to the economy attests to the power and importance of the sector, with total retail sales in the UK constituting around 11 per cent of total economic output (measured as Gross Value Added) in

2016 (Rhodes and Brien, 2017). For every £1 spent in the retail sector (both in shops and online) in 2016, 40p was spent in food stores (ibid.). The retail sector is the largest industrial sector in Great Britain, employing 4.6 million people in 2015; 15.7 per cent of the total. Eight of the top ten UK retailers are involved in selling food, with the food retailing giants of Tesco, Sainsbury's, Asda and Morrisons topping the list (in that order) (Retail Economics, 2018). The 'one-stop' shop of the food superstore, however, is a very recent phenomena, a product of the last 30 years. The revolution gained momentum in the 1960s, an era marked by the emergence of the 'superstore', initially in the north of England by the retailing company Asda (see Box 4.1). The concept spread throughout the country in a series of distinct phases as other companies followed suit, although this diffusion process was by no means temporally or spatially even (Davies and Sparks, 1989). The stimulus to the growth in food retailing is widely attributed to the abolition in 1964 of the RPM, the system whereby goods had to be sold at a price dictated by the food manufacturer (Gurney, 2017; Mercer, 1998). RPM effectively gave food manufacturers considerably more power within the food chain than the retailers. Abolition of the RPM heralded the beginning of price competition, the era of the 'loss leader', to entice customers with a bargain, enabling retailers to achieve a competitive edge. It thus transferred power along the food supply chain to the retailers who were now in a position to determine the price of goods sold.

The growth of the superstore was associated with a marked spatial shift in the location of retailing from city centre to out of town/edge of town. Locating a superstore close to central city/town shopping streets was almost impossible, not just because of the size of the store itself but because of the transport infrastructure and car parking required to service such large shopping centres. In any case, by this stage retailers were persuaded that out of town centres had a huge number of other potential advantages. By drawing

Box 4.1 Defining supermarkets and superstores

Supermarket: A single-level, self-service food store of less than 25,000 sq.ft (or 2,500 sq.m.) sales area. The lower size limit has been variously defined as 5,000 sq.ft, 500 sq.m. or 10,000 sq.ft.

Superstore: A single-level, self-service store offering a wide range of food and non-food merchandise, with at least 25,000 sq.ft (or 2,500 sq.m.) sales area and supported by car parking. The term can also be used to describe non-food stores of the retail warehouse type.

Hypermarket: In Britain, a superstore of at least 50,000 sq.ft or 5,000 sq.m. sales area. In other western European countries, the term is often used to describe superstores.

Source: Guy (1994)

customers away from town centres, the superstores could successfully compete across a wide range of goods other than food.

The increasing size of retailing outlets, associated with the development of the food superstore, has been accompanied by a decline in the overall number of outlets. Further, from the outset, superstores were operated by multiple retailers, defined in the UK as any company operating at least two retail outlets ('small multiple') or at least ten ('large multiple') (Guy, 1994). In most cases, retail companies own many more than ten stores and their rapid expansion has meant a dramatic reduction in the number of independent and co-operative retailers.

Thus the large multiples more than doubled their share of total retail sales (food and non-food) from 22 per cent to 56 per cent in the UK between 1950 and 1982 (Wrigley, 1987).[1] In food retailing the trend was even more dramatic. In 1950, independent grocers accounted for 78 per cent of UK grocery sale but by 1984 their market share had tumbled to below 30 per cent (Wrigley, 1987), and by 2015 to less than 10 per cent (Defra, 2017b). Meanwhile, the growth in market share by the multiple food retailers increased from 23 per cent in 1950 to 57 per cent in 1990 (Burt and Sparks, 1994) and to 93 per cent in 2015 (Defra, 2017b).[2] In terms of floor space, in just the period between 1989 and 1992 in Britain multiple grocers increased their floor area by 5.7 per cent, while the area held by independent and small co-operatives declined by 5 per cent and specialist food retailers by 7 per cent (Langston et al., 1997).

The process has been faster in the UK and elsewhere in northern Europe than in southern Europe. But even in countries with a long tradition of small retail food outlets, such as Italy, the situation is changing. The modern retailing sector's share of food sales in Italy doubled in the 1990s from 20 to 40 per cent (Loseby, 1998) and by 2015 had reached 74 per cent (Global Agricultual Information Network, 2016). By 1994, over 60 per cent of pasta and 46 per cent of olive oil retail sales in Italy were through supermarkets or hypermarkets (Loseby, 1998). Between 1982 and 1995, the number of Italian supermarkets grew from 1,521 to 4,198 (Loseby, 1998) and by 2015 this figure had more than doubled to 8,715 supermarkets and a further 837 hypermarkets (Global Agricultual Information Network, 2016). Figure 4.1 shows how the growth of floor space more than tripled during this period.

In most countries, the number of multiples has declined too as a smaller number of increasingly powerful and highly profitable companies have come to dominate the sector. Table 4.1 shows the changing market shares of the leading UK grocery companies over the period 1970–2017. In recent decades, Tesco has become the clear market leader. The concentration has partly been achieved by a series of substantial mergers and acquisitions of smaller retailing companies (along with many of the sites and stores they occupied). For example, Tesco acquired Hillards in 1987 followed by the Scottish chain Wm Low in 1994; Co-op stores are derived from The Co-operative Group's acquisition of Somerfield in 2008 (formerly Gateway, which was a merger of many companies including International Stores in 1984 and Fine Fare

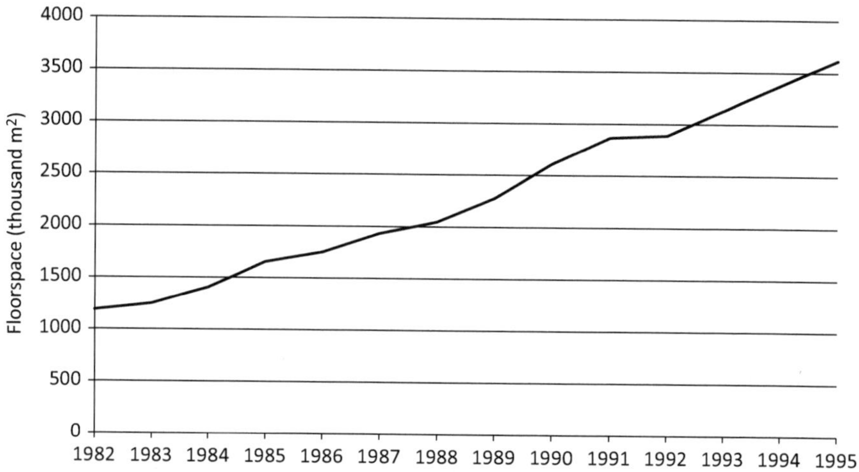

Figure 4.1 The development of supermarkets in Italy, 1982–1995

Source: Loseby, 1998

Table 4.1 Grocery market shares of leading companies in the UK (% total sales)

	1970/71	*1979*	*1990/91*	*2000*	*2010*	*2017*[†]
Tesco	7.2	13.6	12.0	25.0	30.6	27.5
Sainsbury's	6.1	11.9	12.4	17.9	16.6	16.1
Asda	1.5	7.3	8.4	14.1	16.9	15.6
Morrisons*	–	–	1.7	4.9	12	10.4
Aldi	–	–	–	1.5	3.1	6.9
Co-ops/The Co-operative	–	17.4	10.9	5.4	6.5	6.1
Waitrose	–	1.3	2.0	2.7	4.2	5.2
Lidl	–	–	–	1.3	2.4	5.0
Iceland	–	–	1.4	2.8	1.9	2.2
Total	**14.8**	**51.5**	**48.8**	**75.6**	**94.2**	**95.0**

Notes: [†]As at July 2017.
*Morrisons bought Safeway in 2004. Safeway had held 10.4% of the market share in 2000.
Note that the columns in the table illustrate general trends but are not directly comparable as they originate from different sources.

Sources: Smith and Sparks, 1993 (using Akehurst 1984 and Mintel 1992) for 1970–90 data, and Shaw Food Solutions (2017) (using Kantar and IGD) for 2000–2017 data)

Table 4.2 Expansion of multiple retail fascias 1989–1992

	Floor space (1000ft²) Spring 1989	*Floor space (1000ft²) Spring 1992*	*Rate of change %*
Tesco	8,540	10,583	+7.4
Co-op	6,800	7,193	+1.9
Safeway	4,340	6,509	+14.5
Kwik Save	4,380	6,355	+13.2

Source: Langston et al., 1997

Table 4.3 Total UK floor space (thousand ft²) of the 'big 4'

	1995	*2016*
Tesco	13,397	45,253
Sainsbury's	10,801	23,202
Asda	8,436	Unavailable
Safeway/Morrisons*	8,416	14,142

Note: *Morrisons bought Safeway in 2004.

Sources: Data for 1996 are taken from Langston et al., 1998 p.52. Data for 2016 are from Tesco plc, 2017, J Sainsbury plc, 2017 and Morrisons, 2017

in 1986); and Morrisons increased its presence significantly in 2004 with the purchase of Safeway (itself the product of a 1987 merger between a subsidiary of American Safeway and Argyll Foods, which had already acquired Allied Suppliers in 1982 and Hintons in 1984). Nevertheless, 'organic' growth through massive investment in new store development programmes has also assumed increased importance to the major retailers. This culminated in a period of intense competition for prime retail space in the late 1980s, often referred to as the 'store wars' (see Table 4.2).

Many commentators refer to the 'big four' retail corporations in the UK of Tesco, Asda, Sainsbury's and Morrisons as dominating the food retail market (Table 4.3). The food retailing sector has come to be characterised by a polarisation between these 'strong' and the other, 'weaker' multiples, which include Waitrose and Co-op, although the major discount operators have gained considerable market share at the expense of the 'big four' since the financial breakdown in 2007. Aldi is now the fifth largest grocery retailer, and in 2015 the three largest discounters (Aldi, Iceland and Lidl) had a combined market share of 13 per cent (Defra, 2016).

The 'big four' retailers held over 70 per cent of the combined market share in the three months to 31 December 2016 (Kantar Worldpanel, 2017a) and were the top four retailers by sales across the whole UK retail sector in 2015/16 (Retail Economics, 2018). The story of one of these famous names is given in Box 4.2. This concentration of retail capital is more marked in the UK than

Box 4.2 The Sainsbury's story

Sainsbury's origins go back over 125 years to a single grocer's shop in London in 1869. In 1969 it had 89 supermarkets, but by 2017 J Sainsbury plc owned 605 supermarkets and 809 convenience stores, as well as Sainsbury's Bank, which provides financial services (including credit cards, insurance, travel money and personal loans) to 1.8 million active customers. In 2017 Sainsbury's also purchased Home Retail Group plc, which includes Argos and Habitat, further expanding the company's 'offer' to customers so that it now encompasses food, general merchandise (including home, technology, leisure, toys and electrical products), clothing and financial services. Total sales from the Group amounted to £29,112 million in 2017, with a before-tax profit of £581 million.

Like similar companies, online sales have become an important focus for Sainsbury's, growing 7.1 per cent in 2014/15, 8.8 per cent in 2015/16 and 8.2 per cent in 2016/17. The supermarket delivers around 276,000 orders per week (over 14.3 million per year) and expects the demand for their online grocery delivery service in London to double over the next eight years. Customers in some areas can now have their groceries delivered on the same day as ordering and the company is currently trialling one-hour delivery via bicycle to some postcodes in London. Since 2016, customers have also been able to order their groceries via an app, increasing their ability to shop from wherever they may be.

Sainsbury's is, however, facing continuing cost pressures due to increasing prices of raw materials and energy, making cost-saving a key priority. By the end of 2017/18 they will have made £500 million of cost savings across the business over three years, and are planning another £500 million of savings over the next three years. Strategies to achieve this include making checkouts faster and more efficient, reducing spend on property (in favour of digital and IT infrastructure) and using technology to forecast demand to give food extra shelf-life. However, job losses have been seen: 800 store jobs were cut in 2014 and a further 400 in 2017, in addition to 1,000 jobs at its head office (Marlow and Armstrong, 2017).

Food retailers are also coming under increasing pressure to improve the health and nutrition of the food they provide. In 2005 Sainsbury's introduced 'traffic-light labelling' on their own-brand foods and are working to reduce the amount of sugar, salt and fat in their products. For instance, in 2015 they reduced the sugar in their own-brand yoghurts by nearly 20 per cent. They have also increased their range of allergen-free products. The company is keen to promote itself as sourcing ethical and sustainable produce. In 2015/16 Sainsbury's was the largest retailer of Fairtrade products and – in response to consumer demand – are working to increase their provision of British produce, as well as improve animal welfare standards among their suppliers. They are also committed to reducing their packaging, waste and environmental impact.

Source: J Sainsbury plc. (2017)

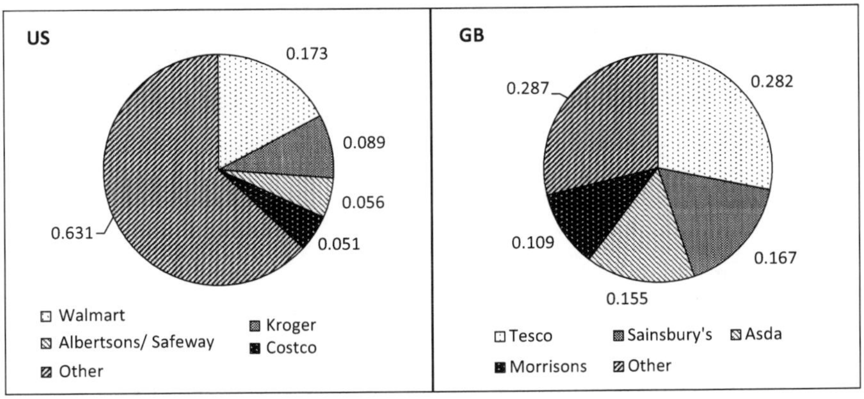

Figure 4.2 Top four retailers by market share, US and GB, 2016

Source: Statista, 2017 (US) and Kantar Worldpanel, 2017 (GB)

elsewhere in the developed world such as the US, where the top four US grocery corporations' combined market share in 2016 was 36.9 per cent (Statista, 2017) (see Figure 4.2). Hughes (1996) points out how the growth of oligopoly in UK retailing allowed successful retailers to generate very high profit margins. All of the larger grocery retailers enjoyed increasing gross and net profit margins during the 1980s (Wrigley, 1991). For example, Sainsbury's net profit margins rose from 4.6 per cent in 1983 to 7.1 per cent in 1989, while Tesco achieved an even greater increase in net profit margins over the same period from 2.7 to 6.1 per cent. These net profit margins were 'quite extraordinary' in international terms (Wrigley, 1991) and can be illustrated by comparing UK retailer's margins in a basket of 11 foods with those of retailers elsewhere in Europe. In the UK these were £7.19, compared to £4.57 in Germany and £6.44 in Denmark. In recent years, however, profit margins have dropped significantly due to intense competition and 'price wars' between the major supermarkets and discounters. Tesco's net profit margins, for instance, dropped from 4.35 per cent in 2012 to an all-time low of -10.35 per cent in 2016, representing a loss of £6.4 billion. Similarly, though less dramatically, Sainsbury's net profit margins dropped from 2.68 per cent in 2012 to a low of -0.7 per cent in 2015, recovering to 1.87 per cent in 2016 (Trading Economics, 2017).

But the retail revolution story is a constantly changing one and the big brand retailers have not had it all their own way. Since the 1990s, as noted earlier, supermarket food prices have come under pressure from discounters such as Kwik Save, Iceland, Aldi and Lidl. Although with 94 per cent of customers who shop at a discounter also shopping at a traditional multiple (Newton Europe, 2017), the likes of Tesco and Sainsbury's are unlikely to disappear any time soon.

Perhaps more importantly, globally, hyper and supermarkets are beginning to lose market share to e-commerce, discounters, convenience stores, cash and carry and traditional formats. Kantar Worldpanel (2017b) predict that by 2021 they will account for just 48 per cent of global spend on FMCG

Table 4.4 Global market shares in FMCGs, 2015–2016 (%)

Channel	*Global value share 2015*	*Global value share 2016*	*Percentage value increase (year-on-year)*
E-commerce	3.80	4.6	26.0
Discounters	5.5	5.6	5.1
Convenience	4.6	4.6	4.1
Cash and carry	1.1	1.4	4.1
Hypermarkets and supermarkets	53.2	52.0	0.7
Traditional	26.1	26.1	3.2
Door to door	0.8	0.8	0.7
Drugstore and pharma	0.6	0.6	0.90

Source: Kantar Worldpanel, 2017b

(fast-moving consumer goods), compared to 53.2 per cent in 2015 (Table 4.4). There is also growth in online sales – accounting for 7.3 per cent of the grocery share in 2016 (making Britain second only to South Korea in terms of the proportion of groceries bought online) (Kantar Worldpanel, 2017c).

How have these phenomenal growth rates in food retailing been achieved? The success of the food retail sector needs to be viewed against a backdrop of general developments within the UK economy, in particular the rise of the service economy, of which retailing forms one significant element. Whereas the manufacturing industry previously dominated, the service sector is now the most significant contributor to the economy. However, this general trend in economic development does not help to explain how food retailers have come to assume such a dominant position within the contemporary geographical and economic landscape. To do this, we must examine a range of operational changes internal to food retailing companies that have permitted growth to occur. These are grouped together and briefly discussed below.

The public face and service of the retailers

Here, store design, location and size, together with the nature and range of products and other customer services, need to be considered.

Increasing store size. The growth in profit margins of the largest food retailers has been underpinned by a growth in the average physical size of retail outlets, associated with the superstore and large supermarkets. Much of this physical expansion has occurred in off-centre locations, where space is available for superstore buildings and the associated infrastructure of car parking. Building new stores is a very costly business. An average cost of developing a typical supermarket store of around 35,000 ft was £25 million in 1991 prices (Wrigley, 1991). Retail companies need to have access to significant sums of

Table 4.5 Own-label share (% value) of the leading four UK supermarket chains

	1980	*1992*	*2012*	*2013*	*2014*
Sainsbury's	54.0	54.0	48.8	49.5	50.7
Tesco	21.0	41.0	48.0	48.2	47.9
Asda	25.0	32.0	46.2	45.8	45.8
Safeway/Morrisons	28.0	35.0	44.6	44.8	45.3

Sources: Hughes (1996, p.2203, using AGB Superpanel data) for 1980–1992. Spary (2014, using Kantar Worldpanel data) for 2012–2014

finance to service their expansion programmes and the levels of profitability they have achieved in recent years has ensured significant levels of capital investment.

Concentration on a single trading format. That is, conformity of shop design in spite of location. The leading multiples have increasingly focused upon a single trading format of 'conforming' superstores with a very broad range of products and services in an effectively standardised format, trading from particular types of location. These conforming stores allow economies of replication and scope and the full development of the retailer as the brand (Burt and Sparks, 1994).

Retailer brand increases. The development of own-label products was adopted as a key competitive strategy in the 1980s by all the major retailers. Table 4.5 shows the increase in own-label shares by the leading retailers. As Hughes (1996) argues, by becoming involved in the production and sale of their own-label commodities, the retailers are potentially able to erode manufacturers' share of the market, by influencing patterns of consumption in their favour and by using their bargaining power to lock manufacturers into retailer-led supply situations for their own-labels. In this way retailers are able to increase their overall profits, particularly if own-labels are in high-margin categories (such as fresh, prepared convenience foods): 'The retailer, to all intents and purposes has become the brand' (Smith and Sparks, 1993, p.43).

Range extension and added value. The continued search for profits by food retailers, combined with shifts in consumer demands, has led to an extension in the range of foods available – a process of 'product differentiation'. This extension has included frozen products in the past but now involves chilled and microwaveable products ('value-adding'), as well as developments in organic and fresh produce.

Product and service quality. During the 1980s, partly in response to changing customer preferences but also as a deliberate competitive strategy, the major food retailers became more reliant on product and service quality to ensure custom, rather than price competition which was favoured as a competitive strategy during the 1970s (i.e. retailers tended to compete with one another on the basis of price, under the banner of 'pile it high and sell it cheap'). Increasingly, supermarkets compete on the basis of convenience and customer service. The Financial Times (16.6.95) reported that Sainsbury's spent £9 million and 18 months on improving its customer service, e.g. by staff retraining (1994–95), following a survey of its shoppers.

Provision of a range of other customer services. During the 1990s, loyalty cards, financial services, home ordering (e.g. via the internet) and home delivery (Cairns, 1996) were all added to the retailer's portfolio in their effort to maintain and increase market share. The UK is the biggest online grocery market in the world after China (Zissis et al., 2017) with a market share of 4.3 per cent in 2014 estimated to rise to 8.3 per cent in 2019 (IGD, 2014). Over the last 10 years, online UK grocery sales have increased by around 17 per cent per annum compared with an annual growth of less than 5 per cent in grocery sales generally (IGD, 2015).

Controlling the food supply chain

Part of the explanation for the growth in food retailing lies in the increasing ability of the large food retailers to exert more control over, or integrate, the food supply chain. Improving efficiency, saving costs and ensuring a consistent supply of quality products, all of which add up to a competitive edge, are all contingent upon the supply and delivery of goods to the retail outlet. The development of food superstores represents the most obvious reflection of the growth and power of food retailers. However, a much less visible, but nonetheless crucial, aspect of this growth relates to the food distribution systems that transport the food from the suppliers to the superstores. The logistics of moving such quantities of food are complex and it is easy to forget or underrate the extraordinary feat of moving all that food, mostly perishable, to consumers (Raven et al., 1995). This is also a clear example of how most farmers are both dependent on, and far removed from, complex systems over which they have little control. The physical distribution systems of the major food retailers have undergone fundamental changes. In particular, they have become increasingly centralised, with a limited number of warehouses supplying retail outlets (as opposed to suppliers delivering direct to stores), and subcontracting to specialised distributors. Investment in information technologies such as EPOS (electronic point of sale) laser scanning has been tied up with the restructuring of the distribution system. Better information flow along the supply chain allows information on sales to be rapidly transmitted to stock control departments, and in turn to suppliers, to ensure availability of goods – thereby improving inventory control, reducing costs and improving competitive edge: 'Technology is being applied throughout the distribution channel to facilitate rapid and accurate information flows and thus timely and appropriate product distribution' (Smith and Sparks, 1993, p.47).

'Backwards' control of the supply chain by retailers represents one element of the tendency towards integration of the food supply chain. Integration is also partially achieved through the development of own-label supply chains in which retailers exert direct control over the manufacturers of own-label products. A further aspect of control and integration of the supply chain is represented by the direct contact between supermarkets and agricultural producers.

The regulatory environment

The exceptional growth in market share and profits achieved by a small number of retail companies has undoubtedly been facilitated by the laissez faire regulatory environment in which these companies operate. This is readily demonstrated by contrasting the UK experience of retail growth with the US. In the US, the retail sector is characterised by relatively lower levels of corporate concentration, profitability and productivity than the UK. Further, whereas the UK retailers tend to be of national significance, in the US this is not the case – food retail companies tend to have a regional presence only. Wrigley (1992) attributes these divergences in the development of the retail sectors in these two countries in terms of differences in the regulation of retailing, specifically the interpretation and use of 'anti-trust' (competition) legislation. From the 1930s to the mid-1980s, US retailing was tightly and strongly regulated by various pieces of competition legislation. In essence, this legislation was hostile to the development of big capital and concentration of market share. In contrast, regulation of the UK retail sector has been 'lenient, pragmatic and benign', permitting the concentration of retail capital and a uniformity in its spatial offering. Changes in US legislation in the late 1980s have brought the American situation in closer alignment with the UK regulatory environment. Nonetheless, the contrasting histories of retail regulation in the two countries have created distinctive patterns of retail growth.

Continuing growth or retail retrenchment in the 1990s?

The rates of physical growth, market share and profit margins outlined above, particularly among the largest retailers, have lead some commentators to describe the 1980s as 'the golden age of retailing' (Wrigley, 1991). However, events in the early 1990s appeared to present a threat to continued growth. The recession of the early 1990s was acutely felt in the property market and service sector, causing a major property crisis for the dominant food retailers. In simple terms, this crisis was an outcome of the rapid growth in retailing during the 1980s where the retailers had become used to paying property costs well above those in other sectors of the British economy. As that economy moved into recession the superstore properties were seen as substantially overvalued and major write-offs were necessary to bring retail property valuations back into line. This property crisis implied new approaches to retaining market share not related to large-scale store expansion programmes alone. Further, increasing environmental concern over the impact of superstore development on green-field sites and the decline in the quality of town centres led to the issuing of new planning guidelines in 1993. This posed a real threat to out of town developments and the opportunities for new superstore development associated with these sites. Wrigley (1998) presents evidence of an increase in the rate of annual planning refusals from less than 50 per cent in the late 1980s to more than 70 per cent after 1993. In addition to the effects of recession and changes in the planning system, some argued that a slowing down of retail development was inevitable given the declining availability of suitable sites and the threat of retail provision outstripping population growth, that is, market saturation.

In spite of these concerns, research by Langston et al. (1997; 1998) suggested considerable *potential* for continued retail growth and buoyancy arising from the enormous geographical variation in retail provision, both regionally and locally, and the spatial variability in regional market share by individual retailer. For example, at that time Sainsbury's and Tesco had a limited presence in Scotland, suggesting an opportunity for these companies to increase their market share in this location. Statements by the major multiples issued in the early 1990s demonstrated an intention to expand in many areas of the country, for example, 150 new Sainsbury stores, 200 new Tesco stores. Moreover, both Tesco and Sainsbury's acquired overseas supermarket chains – Tesco in France and Hungary and Sainsbury's in the USA. Levels of *actual* retail growth during the 1990s (albeit at a slower rate than the 1980s) also tend to contradict the suggestion of retail decline and imply that the impact of Planning Policy Guidance should not be exaggerated. For example, the number of new store openings in the 1990s remained at late 1980s levels, with the exception of Asda who faced debt problems in the early 1990s. Figures of total floor space and numbers of stores 1991–96 also suggest that new retailing development has been far from 'killed off' in the post-property crisis era (see Table 4.3). This growth is not only accounted for by the major retailers, but also by the medium-size to smaller-size players who, in terms of store numbers, account for the greatest levels of expansion.

Within an environment of harsher competition (e.g. by entry into food retailing of limited-line discount retailers) the picture of retail development since the 1990s has been one of continued, if slower, growth and, perhaps more importantly, changes in the form of this growth. These changes can be summarised as follows:

- Following the property crisis, new stores built in the 1990s tended to be smaller, e.g. Tesco have been building 'compact' 16,000–25,000ft^2 stores, as opposed to previous new store developments of the order of 35,000–40,000ft^2. This trend did not imply significantly reduced profits. In fact, lower land and construction costs of these smaller stores, coupled with increased sales densities gave rise to increased rates of return on investment. The trend for smaller stores has continued in the twenty-first century, reflecting a shift in consumer behaviour towards buying less more often in 'top-up' rather than big weekly shops. Thus in 2014, for instance, the amount of floor space 'in the pipeline' for supermarkets dropped from 8.51million ft^2 to just 15.22million ft^2, despite the number of proposals for new stores continuing to rise (Rickard Straus, 2014). Online shopping has also reduced the space needed in stores, and supermarkets are looking to use unproductive floor space in different ways: the inclusion of Argos shops within Sainsbury's megastores is one example of this (Shubber, 2015).

- Partly in response to the 1996 revision of the Planning Policy Guidance Note 6: Town Centres and Retail Developments (PPG6), a new type of 'spatial switching' of retail capital took place, which in some respects mirrored the trend towards out of town development during the 1960s and 1970s; a reinvestment in and reconsideration of high street locations and smaller towns, previously regarded as having catchments too small to be profitable (a reassessment of old locations of profit extraction) (see Wrigley, 1998). This is illustrated by the market entry of the continental European limited-line deep discounters Aldi, Netto (1990) and Lidl (1994), initially into high street locations, as well as the introduction of Tesco's 'Metro' stores in 1993 and Sainsbury's 'Local' stores in 1998.
- An internationalisation of retail investment. Some of the major retail companies have embarked on extensive programmes of overseas investment, permitting continued growth but in foreign locations. For example, the purchase of Shaw's supermarkets in the United States by Sainsbury's in the 1980s (see Wrigley, 1997) and the movement of Tesco into France, Ireland and Eastern Europe in the 1990s (see Palmer, 2005) – and later into Asian countries such as Thailand and Malaysia. Another example is the acquisition of Asda by the US giant Walmart in 1999 – making it part of the world's biggest retailer (ASDA, 2018). Wrigley (1992) argues that this internationalisation of retail capital represents one moment in a series of spatial shifts in the location of retail activity. The foray into international markets has not always been successful for supermarkets, however. For instance, in 2006 Tesco announced plans to expand into the US, only to finally pull out in 2013 having never made a profit there. Sainsbury's also exited the US market in 2004 amid troubles in their UK domestic market, and a venture by them into Egypt in 1999 also proved misguided, resulting in huge losses and their withdrawal from the country in 2001 (Ruddick, 2015).
- The emergence of a higher degree of integration across the supply chain as retailers seek to ensure traceability and increase the information shared with suppliers to ensure visibility across the supply chain, as well as developing strategic relationships to maximise marketing opportunities and enlarge the customer network (Kumar et al., 2017; van Donk et al., 2008).
- The market growth of ready meals, thereby increasing value take for the retailer, with the UK market growing by 35 per cent between 2013

and 2017 (Koch, 2018). A third of the British adult population consume ready-made meals once a week, more than twice the rate than in France (Schmidt Rivera and Azapagic, 2016).

Farmers and the retail revolution

At first sight, it might be expected that farmers would have suffered greatly as the power of retailers increased, and in the concluding section to this chapter we will examine some of the negative reactions to the hegemony of the big retailers and some of the changes, both reformist and radical, that have resulted. But it is important to emphasise, as we do in this section, that for much of the time during which supermarkets came to dominance farmers were protected from some of the worst potential consequences of their unequal terms of trade with the big retailers for reasons rooted in earlier policy concerns. And we will consider some of them here.

We have mentioned already RPM (Resale Price Maintenance), but this was just one of a range of measures that helped farmers in the face of market weakness. From the vantage point of the end of the second decade of the twenty-first century, it is hard sometimes to recall the depth of market (and marketing) intervention throughout most of the twentieth century. Marketing boards, marketing schemes, and direct or indirect price guarantees abounded and these have been described at length elsewhere (Bowler, 1979; Giddings, 1974; Smith, 1990; Winter, 1996). Together they served to shield farmers from the full implications of the changes in retail. Much of the impetus for this was not in price support per se but in market control through the powers given to producers by the Agricultural Marketing Acts 1931 and 1933. The characteristics of marketing schemes varied widely with some much more successful than others. A couple of the more successful, at least in terms of protecting farmers from the full power of retailers within the market place, are discussed here – the Potato Marketing Board and the Milk Marketing Board.

Marsh (1985) admirably summed up the situation with regard to potatoes, and marketing boards more generally, as follows:

> The Agricultural Marketing Acts of 1931 and 1933, which provided the post-war basic for the Potato Marketing Board's (PMB) existence, rested upon the conviction that some regulation of the sale of goods by farmers was necessary to attain stability in markets, reasonable returns for producers and an element of countervailing power vis-a-vis wholesalers and processors. The atomistic structure of the farming sector and the price inelasticity of demand meant that purely voluntary approaches to market control, for example 'by co-operation', could be undermined if a few farmers refused to accept restraint and sought to profit from disciplines exercised by others. Hence, marketing boards were given statutory powers to compel all producers to operate within the framework of an approved marketing scheme. Such arrangements could only be introduced if

> two-thirds of producers, representing at least two-thirds of output, voted for them. They were also dependent upon parliamentary approval and operated under the watchful eye of the Ministry of Agriculture.
>
> (Marsh, 1985, pp.330–331)

Over the years there were numerous twists and turns in the potato marketing story but none fatally undermined the protection for producers in terms of prices. With the outbreak of the Second World War the Ministry of Food assumed control of potato production and the Board's activities were suspended. In 1955 the Potato Marketing Scheme (Approval) Order 1955 revoked the old Scheme and substituted a new – but very similar – one. The Order was further amended in 1962, 1971, 1976, 1985, 1987 and 1990, before abolition in July 1997 when the British Potato Council, stripped of most of the PMB's powers, was established as a successor. The Board sought to maintain prices for producers through a combination of measures:

- Quotas limiting the area a producer can plant;
- Limiting the number of potatoes coming to market by imposing certain standards, mainly related to size;
- Intervention buying, i.e. removing potatoes from market for use as food for livestock or dumping;
- Research and technology developments.

One of the authors (Winter) well recalls when he was working on a farm in the winter of 1977/78 sorting (riddling) and bagging potatoes for direct farm gate sale. His task changed abruptly when, under the intervention buying scheme of the PMB, the potatoes ceased to be sold but were instead loaded into trailers and tipped into a disused railway cutting, a salutary lesson on the downside of state intervention – not only did this appear to be a shameful waste of food but the dumping ground was one of the few remaining sites of some ecological interest on an intensive farm. But this kind of extreme measure clearly protected farmers from extreme market volatility and, arguably, it was the weakness of the PMB not its strength that led to this kind of excess.

The exclusion of small producers, growers with less than 1 hectare of potatoes, from quota requirements combined with annual fluctuations in yield ensured that the protection of price for producers was always an imprecise science. It is hard to argue with Barker's verdict that the PMB was 'only mildly efficient in marketing the produce of its members. It has involved itself mainly with production technology at the expense of any major involvement in the marketing of the commodity' (Barker, 1989, p.150). Moreover, there were issues other than price at work in the potato sector. The area of potatoes grown in Great Britain declined by 58 per cent from 1960 to 2012 and the number of growers fell dramatically and by 2012 was only 3 per cent of the 1960 number (Potato Council, 2012). But production remained almost constant at 6 million tonnes as yields rose, driven mainly by improved crop protection, fertiliser regimes, varieties, and irrigation (ibid). This massive restructuring and the dominance of larger agribusinesses was driven by the need to provide high-quality product to the major processors and supermarkets (Daccache et al., 2012). In particular, it was a trend driven, in part, by supermarket demands for quality, consistency and continuity of supply, which can best be guaranteed by irrigation (Knox et al., 2010).

Turning now to the Milk Marketing Board (MMB), which was formed in 1933, this was a move not primarily designed to provide state support for agriculture – indeed at first many farmers opposed the idea of a Board with statutory rights – but to satisfy the political and economic objectives of a Labour government. In particular, the Board introduced a degree of central economic planning into what socialists of the day considered to be the anarchy and inefficiency of the free market. Competition in the milk sector produced duplications of transport and distribution networks. For example, research showed that some single villages might receive collections from four different milk factories, all sending out lorries to pick up relatively small quantities of milk. When it came to the retailing of milk there were suspicions that the companies acted together to hold up prices to consumers. Not only was this oligopolistic activity seen as artificially inflating of prices, it was also seen as placing a brake on increased consumption of milk. Milk, at this time, epitomised a healthy and sustaining product, essential for the improvement of public health. There were also concerns over the quality of dairy products, with widespread suspicion of adulteration (adding water to milk) and the carrying of tuberculosis. Legislation was already in existence to deal with this latter problem, but it required a more centralised means of organising the dairy industry and of regulating milk production (Cox et al., 1990). The context for this was a rapidly changing market. Dairying had undergone a dramatic transition from the 1860s, when 70 per cent of total milk production was devoted to cheese and butter manufacture, to the 1930s when the figure had dropped to 25 per cent (Taylor, 1976). Foreign competition in cheese and butter, opened up by improved methods of refrigeration and distribution, and improved internal transportation for milk, had encouraged the switch to liquid milk production. Demand for milk increased quite dramatically and the costs of inputs of feedstuffs more slowly, producing a relatively strong position for dairy producers as a whole. But this general picture belied the chaos inherent in a sector in which the farmers' nominal market strength had suffered a fundamental reverse. From producing a finished product, butter or cheese, to be marketed by small retailers often in the same locality, farmers were now selling liquid milk to a dairy company, often in a near monopolistic position for sale in urban markets maybe a hundred miles distant (Winter, 1984). At times when supplies exceeded demand, the market position deteriorated even further. In addition, average prices masked considerable discrepancy between regions and even within the same locality (Taylor, 1976).

Notwithstanding the consumer welfare objectives of the Labour administration which brought it into being, from the very outset concerns were expressed that the consumers' best interests, certainly with regard to price, were not best served by a state-sanctioned monopoly. Press criticisms of the Board for over-pricing milk rumbled on throughout the 1930s, anticipating by half a century some of the criticisms of the Common Agricultural Policy in the 1980s. Ironically perhaps, the Board eventually fell victim to European Union rules because it was seen to offer an unfair advantage to British agricultural producers (Banks and Marsden, 1997). The MMB acted as a major promoter of new techniques such as machine milking, artificial insemination, bulk collection, improved breeding, milk recording, and so on. For many of these improvements major capital investment was required and the Board had its own financial assistance schemes, for example to aid conversion from churn to bulk collection of milk. Post-war dairying, under MMB direction, provided a climate for expansion for those producers able to grasp the opportunity. The conversion from churn to bulk collection of milk, which was completed in England and Wales by 1979, provides one of the most significant examples of the 'modernising' role of the MMB (see Box 4.3). The changeover, promoted by the MMB, to improve its own efficiency and costings, was probably the single most important factor in the growth of specialised production on some farms and the withdrawal from production of others. Of course, it was always the big producers who were most able to take up bulk production. By 1971/72, 32 per cent of producers had bulk tanks, producing 54 per cent of all milk (Baker, 1973). A mixture presented problems of duplication of transport and handling procedures. Pricing and grant policies soon reflected this concern, with the instigation of government grants and Board loans to meet the capital cost of installation of bulk collection facilities, and improved buildings and farm roads for tanker lorries. Premium payments for farm-bulked milk were also offered (Baker, 1973). It was at times a contested policy, which undoubtedly prompted the premature withdrawal from milk production of a large number of older, non-specialised producers who could not justify large capital expenditures for the remaining short period of their working lives.

Furthermore, there are ways in which the Board's policies were to the benefit of the specialised family farmers and small capitalists rather than the larger or better placed capitalist farmers. In the first place, it was the avowed intention of the Board to reduce, if not eliminate, the vast regional variations in milk prices received by farmers prior to its formation. At the Board's inception variation was marked – in June 1934 the average price for a gallon of milk in the South East region was 11.7 old pence, and in the Far Western Region (Devon and Cornwall) only 7.83 old pence (Baker, 1973, p.138). A policy of inter-regional compensation largely eliminated such huge discrepancies. The remoter rural areas that benefited from this equalisation of prices were traditional small farm areas of the pastoral west, whereas those who lost their 'comparative advantage' were the larger producers, often producer-retailers, adjacent to the major conurbations. Such farmers made up a small but highly vocal minority opposed to, or critical of, the Board's activities. Similarly, by its statutory obligation to take the milk of *all* registered producers, those

Box 4.3 Responsibilities of the first Milk Marketing Board

1) The discipline of individual producers by their own collective marketing organisation;
2) The determination of prices at which milk could be sold by producers;
3) The pooling of receipts from the different markets for payment to producers;
4) The encouragement of improvement in the quality of milk;
5) The development of services to assist producers in the production of milk;
6) The improvement of transport and marketing arrangements;
7) The improvement and development of markets for both liquid milk and milk for manufacture;
8) The acceptance of certain (rather weak) safeguards to ensure attention to the interests of government and consumers.

Source: Cox et al. (1990)

family farms in unfavourable locations were relatively shielded from the full impact of market forces. In addition, the specialised family producers also had the benefit of various MMB services vital to improving production. Such services, and the research with them, would almost certainly be the preserve of very large producers in the absence of the Board. Nevertheless, for the majority of specialist milk producers, family farmers or small capitalists, the Board is a success story. Between 1933/34 and 1938/39, milk sales in the Far Western region nearly doubled (Whetham, 1978). Between 1924/25 and 1954/55 sales of milk in the Far Western region increased from 30 to 120 million gallons compared to an increase in the South East from 94 to 131 million gallons (Barnes, 1958). Production shifted from areas close to centres of consumption, such as the South East, to areas better endowed for grass growing such as the Far Western counties of Devon and Cornwall. Thus, greater geographical distances grew up between production and consumption with all the attendant complexities of a modern food chain: processing plants, transport arrangements and retailing networks.

When the UK's MMBs were disbanded in 1994, ending over 60 years of structured formal involvement by the state in the operation of the dairy market, the formal link between the farm gate milk price with the milk's end-use was broken. The sector rapidly moved towards retailer (demand-led) supply chain regulation with farmers experiencing the need for the first time in 60 years to negotiate contracts with buyers and widely varying farm gate prices (Franks and Hauser, 2012). Through the 2000s there have been periodic front-page newspaper stories about milk producers' lower prices and resulting fraught relationships with buyers, including the large supermarkets.

For example, in 2015 the House of Commons Environment, Food and Rural Affairs Committee (2015) examined milk prices in the context of fluctuations in milk price and the resulting pressures on UK dairy farmers, many of whom had consequently left the industry. The main conclusion was as follows:

> The Groceries Code Adjudicator's role, concerning the relationship between direct suppliers and major retailers, is too restricted to be of assistance to the vast majority of dairy producers, as they are indirect suppliers. The Government should urgently consider extending this role to provide more reassurance to farmers. The Adjudicator should also have the power to launch investigations instead of only responding to complaints, and the Government has been slow to provide the Adjudicator with the practical powers required to fine companies which break that code.
>
> (House of Commons Environment Food and Rural Affairs Committee, 2015, p.3)

The GCA is an independent adjudicator to oversee the relationships between the largest UK retailers and their direct suppliers, established in 2013 after an investigation into the UK groceries market in 2008 by the Competition and Markets Commission, which found that some large retailers were transferring excessive risk and unexpected costs to their direct suppliers. Despite this, the government announced in February 2018 that there was not enough evidence to support extending the remit of the GCA.

Responses to multiple retailing: reform, rejection, revolution

The previous section clearly indicates continuing challenges and controversies related to the dominance of retail within the UK food supply chain. The growth of supermarket food retailing may have created many benefits for the consumer such as efficiency, choice, value for money, cleanliness, convenience (long opening hours, easy parking, all food under one roof permitting a 'one-stop shop') and an ever-increasing range of goods on offer. However, the retail revolution also has its dissenters and, while academic work exists, this is also territory for claim and counter claim as the multiples lock horns with pressure groups concerned with a range of issues encompassing environmental quality, food quality, price and social welfare.

In a hard hitting report published by the left leaning think-tank, the Institute of Public Policy Research, in 1995 (Raven et al., 1995), the big retailers are condemned on five counts: increased use of road transport; waste generation; the industrialisation of horticulture; reducing the quality of the built environment in and out of town; and giving the consumer a poor deal. Supermarkets, it is argued, have created an increased demand for personal transport (particularly cars) among consumers. More, longer and car-based shopping trips are now undertaken in the course of food shopping. Supermarkets have also increased the demand for freight transport by transporting goods further. Although centralised distribution may have reduced the number of journeys from distribution centres to stores, the same system has increased the length and number of journeys to these centres by suppliers. Supermarkets have used their considerable political influence in demanding upgraded transport infrastructure, particularly roads. There are environmental and social consequences of this road transport dependency, in the pollution generated by vehicles and the inaccessibility of out of town supermarkets for low-income consumers without cars.

Over-packaging of goods by supermarkets and the subsequent generation of waste is also a concern. The dependency of the supermarkets on long-haul freight, industrial production and extended shelf-life mean that large quantities of packaging are indispensable to them. The economic and environmental costs of waste disposal are not borne by the supermarkets but are passed on to society as a whole.

The negative consequences of the retail revolution for fresh produce growers have received more media attention in recent years, particularly in the context of the farm crisis. While supermarkets are enjoying an increasing share of the fresh produce market (>50 per cent), fruit and vegetable growers complain of: supermarket buyers pressing down the prices they receive for their produce; aggressive and adversarial buyers; vulnerability created by a lack of written contracts and fears of 'delisting'; massive investment costs in providing appropriate packing equipment; payment delays; and a lack of alternative outlets. Because of their monopsonistic buying position, supermarkets can dictate prices and product promotions and growers have little choice but to comply if they wish to ensure sales. The concerns of suppliers about their weakening position in the food supply chain have been highlighted over recent years by various farmer protests at retailer distribution centres and supermarkets, for instance in relation to low milk prices in August 2015 (see BBC News, 2015). As noted in Chapter 3, supermarket-driven farm assurance schemes are also increasing the audit burden for farmers, with many small businesses struggling to meet the costs of complying with multiple schemes (Richards et al., 2013).

Are customers really getting the best deal from supermarkets in terms of price and access to choice of foods? Supermarkets argue that they are providing the customer with value for money. However, evidence presented on a Channel 4 *Dispatches* programme in 2013 (21.01.13) suggested that some foods, particularly fresh fruit and vegetables, are available more cheaply elsewhere. The programme asked mystery shoppers in 32 locations around the country to buy three fruit and vegetable items – broccoli, pears and coriander – at the lowest price available from a large supermarket, supermarket convenience store and an independent trader. Overall for the three items, the supermarket was found to be 12 per cent more expensive, and the supermarket convenience store 35 per cent more expensive, than the independent retailer. The biggest divergence in price was for coriander, which cost 77 per cent more at the supermarket convenience store (and 15 per cent more at the large supermarket) than at the independent seller. The only item that was cheaper at the supermarket than from an independent trader was broccoli, but this was only by 2 per cent and the supermarket convenience store was 21 per cent more expensive (Channel 4 *Dispatches*, 2013). Supermarkets have also been accused of 'duping' customers into spending more than they need to, with a Competition and Markets Authority (2015) response to a super-complaint confirming evidence of misleading offers and confusing pricing and packaging strategies in some cases.

The UK enjoys cheaper food than many other developed countries (although direct comparisons are difficult and inevitably results vary according to the particular groceries examined). For instance, a 2016 Eurostat survey of 440 comparable food and drink items across Europe found that most foods in the UK, including bread, cereals, meat and fish were cheaper than the average, with dairy being the only major food group more expensive (Eurostat, 2017). Nevertheless, retail food prices have risen over the past two decades, with food and non-alcoholic beverage prices rising 11.5 per cent in real terms between 2007 and their peak in February 2014, though reductions since then have reduced that to a real term increase in July 2017 of 4.1 per cent compared to 2007 (Defra, 2017b).

The boom in out of town retailing has been at the expense of the environment, with many ecologically rich sites lost to supermarket development. The disappearance of many green-field sites under retail developments, particularly during the 1980s in part explained tougher new planning restrictions on such development in 1993. The detrimental impact on town-centre retailing of the out of town superstore is further explanation of this changed regulatory environment. The loss of independent retail outlets in town centres not only disadvantages consumers unable to access the convenience and price advantages of out of town stores but can reduce the quality of the built and social environment through a lack of economic activity.

The retailers have responded to many of these criticisms with new initiatives to reform their image and re-position their market positioning. For example, Tidy et al. (2016), highlighting that UK supermarkets have been criticised for operating supply chain relationships based on short-term competitive advantage, call for a move from a purely transactional approach to a relational basis of mutual trust and reliance (see also Free, 2008). Only this way, they claim, can resource efficiencies be achieved, for example to reduce carbon emissions with positive results for suppliers and consumers. They examine how supermarkets have sought to respond to the issue of greenhouse gas emissions through Supplier Relationship Management, including the use of formal Supplier Engagement Programmes. For example,

> Asda is mapping its fresh food supply chain … and have set a target for all fresh, chilled and frozen suppliers to be actively using the Sustain and Save Exchange. The 'Resource Saver' benchmarking tool is already

> claimed to have helped 700 suppliers identify £1.1 million in potential savings. … Tesco have carried out Supplier Carbon Reduction Planning with more than 400 UK dairy farms and is continuing its Carbon Footprinting of products to identify supply chain hotspots. Sainsbury's have 2500 suppliers in Farmer Development Groups and using Supplier Sustainability Scorecards, claiming to have realised 70,500 tonnes of CO_2 savings to date. … less impressive is the fact that neither Tesco nor Asda are able to report on significant carbon reduction impacts as yet. Asda do report that it has completed life cycle assessments for milk, potatoes chicken and egg products that should yield a reduction, but it is too early to measure impact as yet. Likewise, Tesco report working on a web-based 'Footprinter' carbon measurement system which will measure the life-cycle carbon footprint of all 70,000 products sold in the UK and suggest areas for greatest impact. So the signs are present that the groundwork is being done against which future progress can be measured and targets tested
>
> (Tidy et al., 2016, p.3300).

Direct marketing: an old method with a new face

Entirely in opposition to the main trends described so far in this chapter, has been the re-emergence in in recent years of direct selling, that is where producers sell directly to consumers. Direct retailing occurs through a variety of means including farm shops, pick-your-own outlets, roadside stands, box delivery schemes and farmers' markets. There are three elements to the trend towards direct marketing. First, it is perceived by farmers and policy makers concerned with the agricultural industry as a way to boost agricultural incomes by cutting out the 'middle man' and/or by adding value to raw agricultural commodities. Second, changing consumer tastes have led to the emergence of niche markets for organic or local produce attractive to some sections of the consuming public. Third, for many organic producers, the centralised food marketing systems represented by the large supermarkets is antithetical to the principles of the organic movement. Direct marketing represents a decentralised marketing system that is more in sympathy with the philosophy of the organic movement.

Within France, direct marketing is extensive, representing 14 per cent of domestic sales of fresh produce (including imports), 7.6 per cent of total food sales nationally and undertaken by 27 per cent of all farmers (Festing, 1995). In contrast, direct marketing in the UK is relatively weakly developed. In 2003, just 3.3 per cent of farmers in England were undertaking direct sales diversification, though this varied between 2.5 per cent in the North West and 5.7 per cent in the South East (2.7 per cent in the South West) (Keep, 2009). Levels of direct marketing are, however, increasing as farmers seek different ways of augmenting their income and by 2016/17, 9 per cent of farm businesses in England were engaged in the processing/retailing of farm produce as a diversification activity (Defra, 2017a).

Box delivery schemes

Box delivery schemes are an increasingly popular form of direct marketing. In early 1999 the BBC's 'Food Programme' (18.1.99) reported that six years previously only two such schemes were in operation but by 1999 there were 400, distributing food to 30,000 people across the UK. Since then the veg box business has boomed and by 2015 there were over 500 schemes, with the two largest operators (Riverford and Abel & Co) delivering food to around 50,000 customers each (Ethical Consumer, 2016). All schemes are based around the central principle of delivering a box of fresh, seasonal and often organic food (typically of fresh fruit and vegetables, but also of dairy produce, bread or meat) for each subscribing household, either directly to the door or to a central drop-off point. Box schemes may be operated by an individual grower or by a wholesaler or company which buys in produce.

The principle benefits for consumers of box delivery schemes, it is argued, are fresher produce derived from known sources of supply at lower prices than supermarkets. The same BBC programme mentioned above reported that a selection of organic fruit and vegetables cost twice as much from the supermarket as from one box delivery scheme. For producers, box delivery schemes provide a guaranteed market for their produce. The closer links between producers and consumers which box schemes entail are reinforced by the involvement of customers in the farms of box scheme operators. Open days may be held and newsletters produced to keep customers abreast of developments on the farm. Some schemes go further than this and have developed systems where local people support the farm more directly, sharing the financial risks and sometimes working on the farm in return for a regular supply of food. This is known as 'Community Supported Agriculture' (CSA Network UK, 2015).

Farmers' markets

Farmers' markets began in the US over 20 years ago. Today, thousands of American producers sell at more than 8,700 markets (United States Department of Agriculture, 2018), with an annual turnover of $1 billion (Agricultural Marketing Resource Center, 2018). Although a relatively new phenomenon in the UK, farmers' markets now appear to be booming. The first market was held in 1997 in Bath, but by 2010 there were over 750 certified by the National Farmers' Retail and Markets Association (FARMA) across the UK. Typically, the markets have been set up by local councils and are held in city centre locations, although some markets are now being held on farms. Markets are held on a weekly, fortnightly, or monthly basis and sell a range of fresh and processed products including fruit, vegetables, and dairy produce. In spite of the effort and commitment that farmers' markets demand from producers, the principle benefit of this system of retailing is that they completely bypass the supermarket system, that is, they cut out the 'middle man'. Growers can thus determine their own margins on produce and even undercut supermarket prices. Other benefits for producers also exist. Direct contact with consumers enables producers to both offer reassurance on how food is produced and to identify new marketing opportunities.

Of course, farmers' markets are not suited to all producers, for example large arable farms specialising in one or two crops. It is the small and medium-sized family farms (and organic producers in particular) that view farmers' markets as a more profitable way to market their produce. A range of other benefits of farmers' markets have also been identified. As farmers' markets are based upon local farmers and growers (they normally have a restriction that the produce sold is produced or grown within a 30-mile radius of the market), they benefit the local economy by helping to keep local farms in business, as well as potentially strengthening links between farmers and communities. There are also environmental benefits in the form of reduced food miles and reduced waste – both in terms of providing farmers with an outlet to sell excess food, and in terms of minimising food packaging (Bullock, 2000).

Conclusions

For much of agricultural history in the capitalist era of the last three to four centuries, marketing has been a perpetual challenge to farmers. How to store and transport food products, how to limit the economic consequences of fluctuating markets, how to avoid bad deals from inadequate market power. Agricultural pressure groups and agricultural policy in the early twentieth century were a response to the marketing challenge more than to any other factor (Winter, 1996). Consequently, and for what can now be seen as a short historical 'blip', in post-war Britain, markets were highly regulated and farmers assured of market access and market stability. The gathering strength of the big retailers and the emergence of neo-liberal ideology put pay to such security. But now even the big supermarket retailers face challenges from new online retailers. And farmers have fought back through the development of alternative supply chains, by working with retailers in the development of niche or bespoke premium price products, or sometimes, as in the case of liquid milk, through garnering public support for campaigns to name and shame retailers marketing low price milk. Fair trade ideas developed in the very different context of third world producers have been deployed in a first world context. Farmers as social, political and economic actors are inextricably tied to the changing world of retail, their destinies linked both to boardrooms of international retail corporations and to the alternative politics of food sovereignty.

Notes

1 According to Office for National Statistics retail data, large-sized, non-specialised food stores (i.e. supermarkets) accounted for 23 per cent of all retailing sales excluding fuel (in value terms) in 1986. This share increased to 37 per cent in 2017 (Office for National Statistics, 2018).

2 Figures from Kantar Worldpanel place the combined market share of supermarkets at around 95 per cent for 2015, but these are not restricted to foods.

References

Agricultural Marketing Resource Center. (2018). *Farmers' Markets* [Online]. Available at: www.agmrc.org/markets-industries/food/farmers-markets [accessed 27/03/18].

Akehurst, G. (1984). 'Checkout': The analysis of oligopolistic behaviour in the UK grocery retail market. *The Service Industries Journal*, 4(2), 189–242.

ASDA (2018). *Our History* [Online]. Available at: https://corporate.asda.com/our-story/our-history [accessed 20/03/18].

Baker, S. (1973). *Milk to Market*. London: Heinemann.

Banks, J. & Marsden, T. (1997). Reregulating the UK dairy industry: The changing nature of competitive space. *Sociologia Ruralis*, 37(3), 382–404.

Barker, J. (1989). *Agricultural Marketing*. Oxford: Oxford University Press.

Barnes, F. A. (1958). The evolution of the salient patterns of milk production and distribution in England and Wales. *Transactions and Papers (Institute of British Geographers)*, 25, 167–195.

BBC News (2015). *Farmers in fresh protests over supermarket milk prices*. [Online] Available at: www.bbc.co.uk/news/uk-33777075 [accessed 27/03/18].

Bowler, I. (1979). *Government and Agriculture: A spatial perspective*. London: Longman.

Bullock, S. (2000). *The Economic Benefits of Farmers' Markets*. London: Friends of the Earth.

Burt, S. & Sparks, L. (1994). Structural change in grocery retailing in Great Britain: A discount reorientation? *The International Review of Retail, Distribution and Consumer Research*, 4(2), 195–217.

Cairns, S. (1996). Delivering alternatives: Successes and failures of home delivery services for food shopping. *Transport Policy*, 3(4), 155–176.

Channel 4 *Dispatches* (2013). *Supermarket fruit and veg pricier than independent shop/market* [Online]. Available at: www.channel4.com/info/press/news/supermarket-fruit-and-veg-pricier-than-independent-shop-market [accessed 27/03/18].

Competition and Markets Authority (2015). *Pricing Practices in the Groceries Market: Response to a super-complaint made by Which? on 21 April 2015*. London: Competitions and Markets Authority.

Cox, G., Lowe, P. & Winter, M. (1990). The political management of the dairy sector in England and Wales. *In:* Marsden, T. & Little, J. (eds) *Political, Social and Economic Perspectives on the International Food System*. Aldershot: Avebury, 82–111.

CSA Network UK (2015). *Community Supported Agriculture: What is CSA?* [Online]. Available at: https://communitysupportedagriculture.org.uk/what-is-csa/ [accessed 26/03/18].

Daccache, A., Keay, C., Jones, R. J. A., Weatherhead, E. K., Stalham, M. A. & Knox, J. W. (2012). Climate change and land suitability for potato production in England and Wales: Impacts and adaptation. *The Journal of Agricultural Science*, 150(2), 161–177.

Davies, K. & Sparks, L. (1989). The development of superstore retailing in Great Britain 1960–1986: Results from a new database. *Transactions of the Institute of British Geographers*, 14(1), 74–89.

Defra (2012). *Food Statistics Pocketbook 2012*. London: Defra.

Defra (2017a). *Farm Accounts in England – Results from the Farm Business Survey 2016/17*. London: Defra.

Defra (2017b). *Food Statistics Pocketbook 2017*. [Online] Available at: www.gov.uk/government/publications/food-statistics-pocketbook-2017/food-statistics-in-your-pocket-2017-food-chain#uk-grocery-market-shares-2015 [accessed 19/03/18].

Ethical Consumer (2016). *Veg Box Schemes* [Online] Available at: www.ethicalconsumer.org/shoppingethically/sustainablefood/vegboxschemes.aspx [accessed 27/03/18].

Eurostat (2017). *Comparative price levels for food, beverages and tobacco*. [Online] Available at: http://ec.europa.eu/eurostat/statistics-explained/index.php/Comparative_price_levels_for_food,_beverages_and_tobacco [accessed 26/03/18].

Festing (1995). *Direct marketing of fresh produce by small-scale farmers in the UK*. Unpublished PhD thesis, University of London.

Franks, J. & Hauser, S. (2012). Milk prices in a deregulated market. *British Food Journal*, 114(1), 121–142.

Free, C. (2008). Walking the talk? Supply chain accounting and trust among UK supermarkets and suppliers. *Accounting, Organizations and Society*, 33(6), 629–662.

Fulop, C. (1964). *Competition for Consumers: A study of changing channels of distribution*. London: George Allen & Unwin.

Giddings, P. J. (1974). *Marketing Boards and Ministers: A study of agricultural marketing boards as political and administrative instruments*. Farnborough: Saxon House.

Global Agricultural Information Network (2016). *2016 Italian Food Retail and Distribution Sector Report*. Washington: USDA Foreign Agricultural Service.

Gregory, H. (2010). It's a super anniversary: It's 50 years since the first full size self-service supermarket was unveiled in the UK. *The Grocer*.

Gurney, P. (2017). *The Making of Consumer Culture in Modern Britain*. London: Bloomsbury.

Guy, C. (1994). *The Retail Development Process*. London: Routledge.

House of Commons Environment Food and Rural Affairs Committee (2015). *Dairy Prices*. Fifth report of session 2014–15. HC817. London: The Stationery Office Limited.

Hughes, A. (1996). Retail restructuring and the strategic significance of food retailers' own-labels: A UK—USA comparison. *Environment and Planning A: Economy and Space*, 28(12), 2201–2226.

IGD (2014). UK food and grocery evolution 2014–2019. Available: www.igd.com/our-expertise/Retail/retail-outlook/21115/The-next-five-years-How-the-UK-grocery-market-will-evolve/ [accessed 02/04/16].

IGD (2015). UK grocery retailing. Available: www.igd.com/articles/article-viewer/t/uk-grocery-retailing/i/15513 [accessed 20/03/18].

J Sainsbury plc (2017). *Annual Report and Financial Statements 2017*. London: J Sainsbury plc.

Jeffreys, J. B. (1954). *Retail Trading in Britain 1850–1950*. Cambridge: Cambridge University Press.

Kantar Worldpanel (2017a). *Grocery market share* [Online]. Available at: www.kantarworldpanel.com/en/grocery-market-share/great-britain [accessed 22/08/17].

Kantar Worldpanel (2017b). *Hyper and supermarkets lose market share globally* [Online]. Available at: www.kantarworldpanel.com/global/News/Hyper-and-supermarkets-lose-market-share-globally [accessed 22/08/17].

Kantar Worldpanel (2017c). *UK online grocery sales reach 7.3% market share* [Online]. Available at: www.kantarworldpanel.com/en/PR/UK-online-grocery-sales-reach-73-market-share- [accessed 15/08/17].

Keep, M. (2009). *Farming Diversification in England: Statistics*. London: Defra.

Knox, J. W., Rodriguez-Diaz, J. A., Weatherhead, E. K. & Kay, M. G. (2010). Development of a water-use strategy for horticulture in England and Wales – a case study. *The Journal of Horticultural Science and Biotechnology*, 85(2), 89–93.

Koch, S. (2018). Trends in food retail: The supermarket and beyond. *In:* Lebesco, K. & Naccarton, P. (eds) *The Bloomsbury Handbook of Food and Popular Culture*. London: Bloomsbury, 111–123.

Kumar, V., Chibuzo, E. N., Garza-Reyes, J. A., Kumari, A., Rocha-Lona, L. & Lopez-Torres, G. C. (2017). The impact of supply chain integration on performance: Evidence from the UK food sector. *Procedia Manufacturing*, 11, 814–821.

Langston, P., Clarke, G. P. & Clarke, D. B. (1997). Retail saturation, retail location, and retail competition: An analysis of British grocery retailing. *Environment and Planning A: Economy and Space*, 29(1), 77–104.

Langston, P., Clarke, G. P. & Clarke, D. B. (1998). Retail saturation: The debate in the mid-1990s. *Environment and Planning A: Economy and Space*, 30(1), 49–66.

Loseby, M. (1998). The impact of the Uruguay Round on the agro-food sector and the rural environment in Italy. *In:* Antle, J. M., Lekakis, J. N. & Zanias, G. P. (eds) *Agriculture, Trade and the Environment*. Cheltenham: Edward Elgar, 170–184.

Marsh, J. S. (1985). Economics, politics and potatoes – The changing role of the Potato Marketing Board in Great Britain. *Journal of Agricultural Economics*, 36(3), 325–343.

Mathias, P. (1967). *Retailing Revolution*. London: Longmans, Green and Company.

Mercer, H. (1998). The abolition of Resale Price Maintenance in Britain in 1964: A turning point for British manufacturers? *Economic History working papers, 39/98*. London: London School of Economics and Political Science.

Mintel (1992). *Food Retailing*. Mintel Retail Intelligence.

Morrisons (2017). *Preliminary Results: 52 weeks to 29 January 2017*. Bradford: Morrisons.

Newton Europe (2017). *Beating the Discounters: Myth-busting and the £4 billion opportunity in grocery*. Oxfordshire: Newton Europe.

Office for National Statistics (2018). *Retail sales pounds data*. [Online] Available at: www.ons.gov.uk/businessindustryandtrade/retailindustry/datasets/poundsdata totalretailsales [accessed 19/03/18].

Palmer, M. (2005). Retail multinational learning: A case study of Tesco. *International Journal of Retail & Distribution Management*, 33(1), 23–48.

Potato Council (2012). *Production and Price Trends 1960–2011*. Kenilworth: Agriculture and Horticulture Development Board.

Raven, H., Lang, T. & Dumonteil, C. (1995). *Off our Trolleys? Food Retailing and the Hypermarket Economy*. London: Institute for Public Policy Research.

Retail Economics (2018). *Top 10 UK Retailers* [Online]. Available at: www.retaileconomics.co.uk/top10-retailers.asp [accessed 19/03/18].

Rhodes, C. & Brien, P. (2017). *Briefing Paper No. 06186. The Retail Industry: Statistics and policy*. London: House of Commons Library.

Richards, C., Bjørkhaug, H., Lawrence, G. & Hickman, E. (2013). Retailer-driven agricultural restructuring—Australia, the UK and Norway in comparison. *Agriculture and Human Values*, 30(2), 235–245.

Rickard Straus, R. (2014). End of the road for supermarket megastores? Retailers plan smaller sites as households ditch big weekly shop. *This is Money.co.uk* [Online]. Available at: www.thisismoney.co.uk/money/news/article-2715679/Death-supermarket-megastore-Plans-new-sites-drop-lowest-level-financial-crisis-households-ditch-big-weekly-shop.html [accessed 20/03/18].

Ruddick, G. (2015). Why do British retailers get it so wrong overseas? *The Telegraph* [Online]. Available at: www.telegraph.co.uk/finance/newsbysector/retailandconsumer/11575184/After-Sainsburys-Egyptian-drama-why-do-British-retailers-get-it-so-wrong-overseas.html [accessed 20/03/18].

Schmidt Rivera, X. C. & Azapagic, A. (2016). Life cycle costs and environmental impacts of production and consumption of ready and home-made meals. *Journal of Cleaner Production*, 112, 214–228.

Shaw, G., Bailey, A., Alexander, A., Nell, D. & Hamlett, J. (2012). The coming of the supermarket: The processes and consequences of transplanting American know-how into Britain. *In:* Jessen, R. & Langer, L. (eds) *Transformations of Retailing in Europe After 1945*. London: Ashgate, 35–53.

Shaw, G., Curth, L. & Alexander, A. (2004). Selling Self-Service and the Supermarket: The Americanisation of food retailing in Britain, 1945–60. *Business History*, 46(4), 568–582.

Shaw Food Solutions (2017). *The Food Desert website: UK grocery market share* [Online]. Available at: www.fooddeserts.org/images/supshare.htm [accessed 22/08/17].

Shubber, K. (2015). Supermarkets battle to mop up excess space. *Financial Times* [Online]. Available at: www.ft.com/content/a94436d2-05fe-11e5-b676-00144feabdc0 [accessed 20/03/18].

Smith, D. L. G. & Sparks, L. (1993). The transformation of physical distribution in retailing: The example of Tesco plc. *The International Review of Retail, Distribution and Consumer Research*, 3(1) 35–64.

Smith, M. J. (1990). *The Politics of Agricultural Support in Britain*. Dartmouth: Aldershot.

Spary, S. (2014). Own label category report 2014. *The Grocer* [Online]. Available at: www.thegrocer.co.uk/reports/category-reports/own-label-category-report-2014/356584.article [accessed 22/08/17].

Stanton, J. L. (2018). A brief history of food retail. *British Food Journal*, 120(1), 172–180.

Statista (2017). *Market share of U.S. food and beverage purchases in 2016, by company* [Online]. Available at: www.statista.com/statistics/240481/food-market-share-of-the-leading-food-retailers-of-north-america/ [accessed 22/08/17].

Taylor, D. (1976). The English dairy industry, 1860–1930. *The Economic History Review*, 29(4), 585–601.

Tesco plc (2017). *Annual Report and Financial Statements 2016*. Welwyn Garden City: Tesco plc.

Tidy, M., Wang, X. & Hall, M. (2016). The role of Supplier Relationship Management in reducing Greenhouse Gas emissions from food supply chains: Supplier engagement in the UK supermarket sector. *Journal of Cleaner Production*, 112, 3294–3305.

Trading Economics. (2017). *Corporate profits* [Online]. Available at: https://tradingeconomics.com/tsco:ln:net-profit-margin [accessed 22/08/17].

United States Department of Agriculture (2018). *Local Food Directories: National Farmers Market Directory* [Online]. Available at: www.ams.usda.gov/local-food-directories/farmersmarkets [accessed 27/03/18].

van Donk, D. P., Akkerman, R. & van der Vaart, T. (2008). Opportunities and realities of supply chain integration: The case of food manufacturers. *British Food Journal*, 110, 218–235.

Whetham, E. (1978). *The Agrarian History of England and Wales, Vol. VIII 1914–1939*. Cambridge: Cambridge University Press.

Winter, M. (1984). Corporatism and agriculture in the UK: The case of the Milk Marketing Board. *Sociologia Ruralis*, 24(2), 106–119.

Winter, M. (1996). *Rural Politics: Policies for agriculture, forestry and the environment*. London: Routledge.

Wrigley, N. (1987). The concentration of capital in UK grocery retailing. *Environment and Planning A: Economy and Space*, 19(10), 1283–1288.

Wrigley, N. (1991). Is the 'golden age' of British grocery retailing at a watershed? *Environment and Planning A: Economy and Space*, 23(11), 1537–1544.

Wrigley, N. (1992). Antitrust regulation and the restructuring of grocery retailing in Britain and the USA. *Environment and Planning A: Economy and Space*, 24(5), 727–749.

Wrigley, N. (1997). British food retail capital in the USA – Part 1: Sainsbury and the Shaw's experience. *International Journal of Retail & Distribution Management*, 25(1), 7–21.

Wrigley, N. (1998). PPG6 and the contemporary UK food store development dynamic. *British Food Journal*, 100, 154–161.

Zissis, D., Aktas, E. & Bourlakis, M. (2017). A new process model for urban transport of food in the UK. *Transportation Research Procedia*, 22, 588–597.

5 Farmers and sustainable intensification

Introduction

Sustainable intensification (SI) was introduced briefly in Chapter 2. In this chapter, we expand on those introductory comments in four ways. First we explore some of the definitional and conceptual issues raised in Chapter 2, including the various criticisms of SI that have emerged alongside the insights from those who have promoted the concept. Second, in taking forward, on this occasion, a *narrow* conceptualisation of SI we explore in greater detail the particular challenges that SI presents to mainstream agriculture in the UK. Third, we look at the evidence from our own research exploring some of the attitudes and responses of farmers to SI. As indicated in Chapter 1, we have used farm survey data gathered from a specific element in the Defra-funded Sustainable Intensification Research Platform[1] (SIP), and this provides the bulk of the empirical data presented in this section. We conclude with some comments on the challenge of adequately capturing the social aspects of sustainability within SI research.

SI: definitional and conceptual issues

It is a feature of our times that academic activity and policy ideas tend to cluster around key phrases which are deployed to imply something radically new and developmental; a paradigmatic shift that perhaps belies the incremental change that, in reality, characterises much of science. Sometimes these shifts occur with startling speed, arguably sacrificing conceptual clarity and methodological rigour along the way. The speed with which both academic and policy communities embraced first 'multifunctional land use', then 'ecosystem services' and more recently 'natural capital' is a case in point, which will be familiar to many who have spent some years studying farming and rural land use. This is not to suggest that these three particular concepts are identical – they are not – but equally nor is it possible, we believe, to claim that they necessarily represent a logical progression towards a more helpful, precise and rigorous conceptualisation, or towards better real-world outcomes. So does SI represent a paradigm shift? Is it complementary to ecosystem services or natural capital or something quite different? There is considerable overlap between the substantive content of seeking sustainable agriculture and seeking the protection and enhancement of natural capital. Lest there be any doubt of this take a look at the nature and breadth of the data layers that were deployed in the development of the SIP Landscape Typology Tool (see Anthony et al., 2016; Henrys et al., 2016).

The power of the term SI is because it is arresting and provocative. It appears to offer solutions to intractable problems by combining seemingly opposing and contradictory elements, sustainability and intensification. And, we would suggest that it has been embraced by some in government precisely because, linguistically at least, it seems to offer resolution and reconciliation between the competing demands for more food and a better natural environment. In that sense, although its origins lie clearly in scientific work, it is also an ideological and political construct. Its oxymoronic quality, decried by some, is deliberate – almost a dialectical approach aimed at solving some of society's most 'wicked' of problems. The same applies, incidentally, to the case of natural capital, which we look at in more detail in Chapter 8, where the political imperative is to resolve the competing pressures of nature and economic development. In the case of natural capital, the underlying mechanism to solve the societal challenge of managing the natural world, in the context of the ever-increasing demands made on it, is to deploy the tools of economics. By contrast, in the case of SI, resolution is sought more through science. Of course, there is an important place in SI for economics and farm businesses management economics is very much at the heart of SI. At its simplest, SI is about efficiency, an economic concept, but while this can be, and often is, measured solely in financial accounting terms, it is *agronomic efficiency* – the ability to deliver the greatest level of food output for the lowest amount of input – that drives much of SI thinking through, for example, closing yield gaps (Mueller et al., 2012) or applying agro-ecological understanding to farming systems to reduce inputs. Thus SI has often been framed as 'producing more from less' with the heavy lifting to be done by agricultural and ecological science.

Like many who confronted the notion of SI for the first time in the late 2000s, we were eager to engage with its possibilities but to do so in the context of broadening its definition away from a narrow notion of SI paying little attention to wider sustainability criteria. The following examples illustrate this broadening:

> A virtue of the relatively new language of ecosystems is that it seeks to place the provisioning services of nature, e.g. food and energy which are produced and sold through market-based processes, on the same basis as the supporting, regulating and cultural services, which are non-marketed. A correct interpretation of sustainable intensification should embrace examples where the output or production which is intensified per hectare

> are the conservation outputs, e.g. pollinators or fledged lapwings per hectare … So pursuing intensification of environmental services per unit of land is critical, and sustainable intensification must put the task of producing non-provisioning ecosystem services alongside the provisioning services of food and energy.
>
> (Buckwell et al., 2014, pp.3–4)

> In its current use, the term 'sustainable intensification' is often weakly and narrowly defined, and lacks engagement with key principles of sustainability. Without specific regard for equitable distribution and individual empowerment (distributive and procedural justice), agricultural intensification cannot legitimately claim to be 'sustainable' nor does agricultural intensification address issues of food security. Food security can be achieved only through a holistic agenda that looks beyond production, targets appropriate spatial and temporal scales, and considers regional conditions.
>
> (Loos et al., 2014, p.356)

We were very much part of that drive to broaden the agenda and to ensure that ecosystem services and the multifunctionality of land use would not be lost sight of, indeed would be central to the promotion and exploration of SI (Fish et al., 2014; Gunton et al., 2016). We still believe that to be the case, in the sense that there is no place in contemporary agriculture for a narrow focus on food production and, within that, solely on agronomic and economic efficiency. SI needs to cover everything that might be expected from land and its management, and this will encompass multifunctionality and the provision of ecosystem services (food and fibre after all are provisioning services in that framework), and maybe too the promotion of nutrition security. However, our views have been modified as we feel that the attempt to broaden must not be allowed to strip SI of any heuristic value. It is important that any broadening is seen as an issue of *scale* as Gunton et al. (2016) have attempted to set out. As we set out in more detail in the next section, we have come to the view that SI should be treated as a middle-level concept and a sector specific concept. To say that SI is about farming is not to suggest that wider global concerns don't matter. Indeed, in terms of global ethics and justice they matter much more than farming per se. But we need tools and interventions that are appropriate and efficacious at the levels in which they are deployed. While it is entirely appropriate for farming to take place in regulatory and deliberative environments that ensure many wider societal requirements are brought into play, farmers work the land, and we believe SI is only really a useful concept if it is applied to the practices of farming. We agree with Cook et al. (2015) that SI is a useful guiding framework for raising agricultural productivity on existing land in a sustainable manner but is not in and of itself a paradigm for achieving food security overall, being just one element of a sustainable food system situated in a green economy.

As far as we have been able to ascertain, the first published use of the term SI was in 1995 by Reardon et al. (1995; see also Clay et al., 1998; Reardon et al., 1997), despite the common assertion that the term was coined first by Pretty (for example, Kuyper and Struik, 2014). Pretty, in his 1997 papers, cites some earlier sources (and does not mention Reardon) but of the sources used by Pretty that we were able to track down, none talk specifically of SI. Hazell (1995) talks not of SI but 'appropriate intensification' and McCalla (1995) of 'environmentally-sustainable production systems'. Even Pretty (1995) himself does not appear to use the term. Reardon's characterisation is very helpful:

> Intensification to date in Africa has meant the use of more labor, shorter fallow times, and denser planting & often without accompanying investments in land conservation and soil fertility. Without sufficient use of fertilizer and organic matter, intensification of land use causes soil erosion and loss of fertility. To break this vicious circle, African farmers need to pursue 'sustainable intensification'. This means using inputs and capital which provide net gains in productivity, but which also protect land and water, and enhance soil fertility over time. Specifically, farmers need to increase the use of fertilizer, lime, mulch, manure, and, in some areas, animal traction combined with tied ridging. They will need to adopt soil conservation investments such as alley cropping, bunds, windbreaks, and terraces. Introduction of perennial crops or integration of forestry/fruit trees/livestock with cropping are other ways of ensuring that intensive farming is sustainable.
>
> (Reardon et al., 1995, p.1)

More recently, Rockström et al. (2017) have proposed that a paradigm shift towards SI translates to these key operational strategies:

- Plan and implement farm-level practices in the context of cross-scale interactions with catchments, biomes, and the landscape as a whole. Maximise farm-level productivity by maximising ecological functions, from moisture feedback to disease abatement, across scales;
- Integrate ecosystem-based strategies with practical farm practices, where natural capital (soil, biodiversity, nutrients, water) and multifunctional ecosystems are used as tools to develop productive and resilient farming systems;
- Develop system-based farming practices that integrate land, water, nutrient, livestock, and crop management;
- Utilise crop varieties and livestock breeds with a high ratio of productivity to use of externally and internally derived inputs;
- Adopt circular approaches to managing natural resources (e.g. nutrient recycling) and mixing organic and inorganic sources of nutrients;
- Harness agro-ecological processes such as nutrient cycling, biological nitrogen fixation, allelopathy, predation, and parasitism;

- Assist farmers in overcoming immediate SIA [sustainable intensification of agriculture] adoption barriers and build incentives for their sustained adoption, rendering the ecological approach profitable in the long run;
- Build robust institutions of small farmers, led especially by women, which enable an equitable interface with both markets and government.

(Rockström et al., 2017, p.9)

SI: pushing the agenda on farms in the UK

So what might this mean for UK farmers? Farmers after all are the 'actors' with 'agency' in terms of just how it is that farmland is managed within the territory of their jurisdiction. When much of the debate on SI is constructed, as already indicated, in terms of societal, even global, requirements, it is sometimes forgotten that it is in private spaces (see Winter et al., 2011) that farming activity takes place and that within such spaces the farming style, defined as 'a distinctive way of ordering the many socio-material interrelations involved in farming' (ibid., p.23) will vary, and consequently so too will SI pathways and outputs.

In this section we explore the specific challenge of SI in the UK. The environmental sustainability issues are explored in greater detail in Chapter 8 so here our main focus is on the agricultural challenges, and in particular the question of productivity, starting with the notions of the yield plateau and yield gap.

As noted in Chapter 3 (see Figure 3.9), the total factor productivity of UK agriculture rose dramatically between the 1970s and 2000s (and in the 20 years prior to the current data series; see Zayed, 2016), but progress has somewhat stagnated since 2005 and there is some concern that growth has fallen behind that of other developed nations (NFU, 2015). In the context of SI, it is also important to note that improvements in total factor productivity since the mid-1980s have largely been driven by falls in the volume of inputs (particularly labour, energy and fertilisers) rather than increased agricultural output.

A particular concern is what has been referred to as the 'yield plateau'; a stagnation in yield advances in key crops, particularly wheat, both in the UK and across Europe since the mid-1990s (Brisson et al., 2010). Figure 5.1 shows how, beyond expected year-to-year fluctuations, there were steady increases in the trend of most cereal yields between the post-war years and the mid-1990s, but that improvements tailed off after this (although there is some indication that yields may now be on the rise again). This yield plateau has prompted concern within the industry and stimulated heightened efforts to enhance yields through both scientific and agronomic improvements. SI is a key response to the yield gap issue.

Genetic improvement of crop varieties has a big part to play in driving productivity improvements. Through a reanalysis of historical series of UK

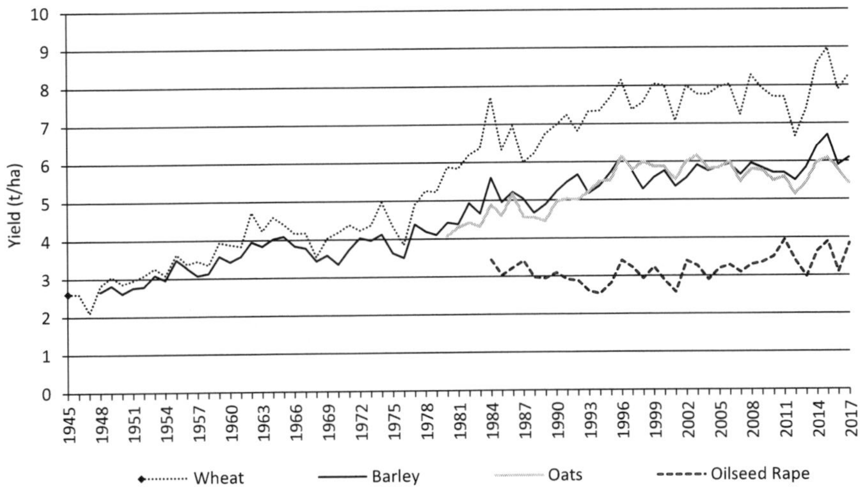

Figure 5.1 Average yields of cereals in the UK, 1945–2017

Source: Defra, 2018

crop variety trials, Mackay et al. (2011) estimate that prior to 1982 at least 88 per cent of yield improvements in wheat was attributable to genetic improvement rather than changes in agronomy. Knight et al. (2012) also report that – despite the yield plateau – wheat yield potential in the UK increased by an average of 0.05 tonnes per hectare per year between 1980 and 2011 thanks to genetic improvements. In seeking greater productivity, the challenge is thus not just to improve genetics through plant breeding, but also to close what is known as the 'yield gap'; that is, the difference between potential yield (the highest theoretical yield given optimal agronomic practices and conditions in a given area) and farm yield (what is achieved in practice). The yield gap is closed when known innovations are adopted faster than new ones are invented and when crop management is improved (Fischer, 2015). It is also important to consider that new varieties can have different nutritional and agronomic requirements, with implications for the application of inputs (and the environmental considerations associated with these). For instance, modern wheat varieties can require as much as 20 kg more N per tonne of grain than older varieties (Sylvester-Bradley et al., 2008). Fischer et al. (2014) estimated the wheat yield gap in the UK in 2010 to be 2.7 tonnes per hectare (34 per cent of farm yield) and noted a trend towards a widening gap. According to the Yield Enhancement Network (YEN), average UK cereal yields are currently around 8 tonnes per hectare but research trials and leading farms often achieve 16 tonnes per hectare and the maximum biophysical potential yield (i.e. under optimal conditions of light energy, rainfall and soil water storage) has been calculated at around 20 tonnes per hectare (YEN, 2018).

Table 5.1 Factors contributing to the national wheat yield trend, 1980–2011 (UK)

<table>
<tr><th>Period</th><th>1980–1996</th><th>1996–2002</th><th>2002–2011</th></tr>
<tr><td colspan="4">Yield gain + or loss – with estimate of t/ha per year. 0 neutral, ( ) uncertain</td></tr>
<tr><td>Farm yield</td><td>+0.105</td><td colspan="2">+0.016</td></tr>
<tr><td>Genetic potential</td><td>+0.05</td><td colspan="2">+0.05</td></tr>
<tr><td>Variety choice</td><td>0</td><td>−0.01</td><td>0</td></tr>
<tr><td>Agronomic effects</td><td></td><td></td><td></td></tr>
<tr><td>Nutrition N</td><td>−0.006</td><td>−0.006</td><td>−0.006</td></tr>
<tr><td>Nutrition S</td><td>(−)</td><td>+0.025</td><td>0</td></tr>
<tr><td>Disease control</td><td>(+)</td><td>−0.01</td><td>+0.01</td></tr>
<tr><td>Rotation</td><td>+0.015</td><td>0</td><td></td></tr>
<tr><td>Cultivation</td><td>+0.001 to +0.002</td><td colspan="2">−0.004 to −0.007</td></tr>
<tr><td>Sowing date</td><td>+0.003</td><td colspan="2">+0.003</td></tr>
<tr><td>Rising CO_2 levels</td><td>+0.011</td><td>+0.011</td><td></td></tr>
<tr><td>Weather and/or other factors</td><td>+0.030 to +0.031</td><td>−0.040 to −0.043</td><td>−0.045 to −0.048</td></tr>
</table>

Source: Reprinted from Knight et al., 2012, with permission from the Agriculture and Horticulture Development Board (AHDB)

In a study for Defra and the HGCA (the cereals and oilseed division of the Agriculture and Horticulture Development Board), Knight et al. (2012) analysed a number of factors contributing to yield changes in wheat and oilseed rape between 1980 and 2011 and, in particular, the yield plateaus over this period (see Tables 5.1 and 5.2). They report that, for wheat, although new varieties, improved crop protection measures and a move to earlier sowing have contributed positively to yield, overall yield increases have been hampered by negative effects of weather conditions, crop nutrition (specifically sub-optimal applications of nitrogen and deficiencies in sulphur in some areas), and possibly by the transition to reduced tillage (though this may be a short-term impact of the transition to this practice) and long-term impacts from deep soil compaction. Similar factors (namely a decrease in N application and increasing S deficiency, an increase in spring oilseed rape, weather conditions, and poor uptake of higher-yielding varieties) also contributed to declines in oilseed rape yields between 1984 and 1994, although yields improved after 2004 thanks to better uptake of new varieties, improved crop protection and more favourable weather conditions (although a shift to shallow cultivation was noted to be detrimental). The authors make a number of suggestions for overcoming agronomic limitations to yields but, importantly, note that the plateau has partly occurred as a result of producers focusing their efforts on maximising profit (through efficiencies etc.) rather than yields per se (Knight et al., 2012). This is entirely justifiable from a business point of view but potentially poses a challenge to the sustainable

Table 5.2 Factors contributing to the national oilseed rape yield trend, 1984–2011 (UK)

Period	*1984–1994*	*1994–2004*	*2004–2011*
Yield gain + or loss – with estimate of t/ha per year. 0 neutral, ( ) uncertain			
Farm yield	−0.04	0.022	0.075
Genetic potential	0.048	0.048	0.048
Variety choice	−0.031	−0.038	0.014
Spring oilseed rape area	−0.009	−0.003	0.008
Agronomic effects			
Nutrition N	−0.02	0	0
Nutrition S	(−)	0.027	0
Crop protection	(−)	0	(+)
Rotation	−0.004	−0.004	
Cultivation	0.003	−0.006	
Weather and/or other factors	−0.027	−0.002	0.015

Source: Reprinted from Knight et al., 2012, with permission from the Agriculture and Horticulture Development Board (AHDB)

intensification agenda and highlights the importance of considering the role, aims and needs of the farmer alongside any technological approaches to increasing food production.

As noted in Chapter 3, productivity increases are, to some extent, associated with factors outside of the individual farmer's control, such as advances in biotechnology, agro-chemical inputs, and machinery. Ultimately, however, it is the farmer's own willingness and ability to implement new technologies and techniques that ultimately determines the extent to which productivity potential is fulfilled. Consequently, as we discuss further in Chapter 6 (see in particular Figure 6.8), economic performance and return on capital employed (an indicator of efficiency and productivity) is highly variable from farm to farm, even among those of a similar size. As we can see in Figure 5.2, high performing farms (those in the top 25 per cent of farms by economic performance) obtained notably greater yields than low performing farms across all crops in 2016/17 and, as Figure 5.3 shows, yields are not related to farm size, with the smallest farms obtaining a similar (or even greater) yield per hectare than the largest farms. A farm's ability to obtain maximum yield will be related in part to their ability to invest in labour and machinery, but management choices can also have an important influence. For instance, in relation to wheat and oilseed rape, Knight et al. (2012) note that in some cases agronomic improvements can be made regarding the appropriate selection of crop varieties, application of N and S, and choices around crop rotation and cultivation timing.

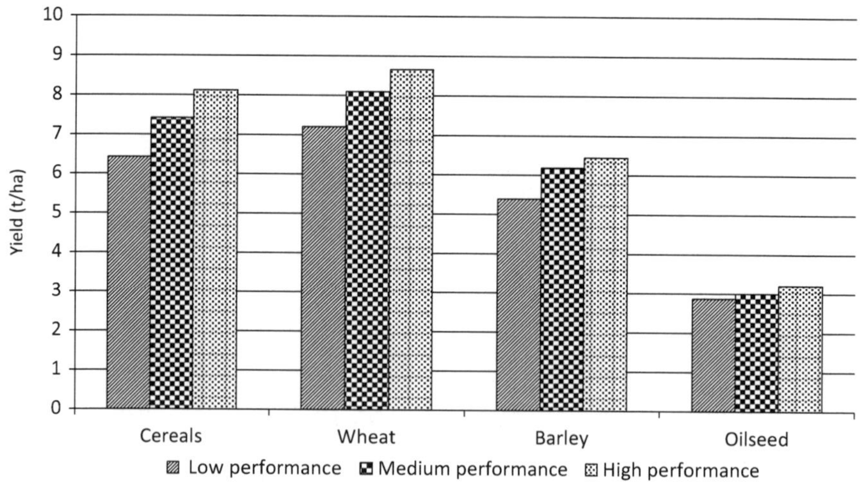

Figure 5.2 Crop yields by economic performance, 2016/17

Source: Defra & Rural Business Research, FBS, 2018

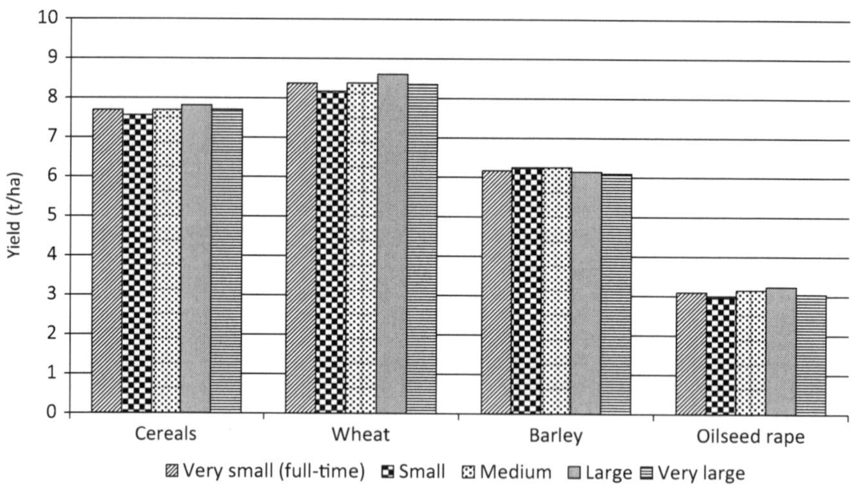

Figure 5.3 Crop yields by farm size (SO), 2016/17

Source: Defra & Rural Business Research, FBS, 2018

SI: what do farmers make of SI?

During the SIP, as introduced in Chapter 1, we undertook a face-to-face survey of farmers within seven case study areas (see Figure 1.1 in Chapter 1). The survey involved face-to-face interviews with 244 land managers (approximately 35 in each case study area) and was designed to collect information on farm type, enterprise structure, size, tenure, employment, formal and informal agri-environmental management, diversification (and associated employment), profitability, and a range of social factors (e.g. health and well-being, community participation and farmers' views and experiences regarding SI).

The SIP Baseline Survey, conducted in 2015, found that 51 per cent of farmers had heard of SI, 41 per cent had not, and 8 per cent were unsure. These findings roughly align with those of a study two years earlier by Firbank et al. (2013), where 50 per cent of farmers in focus groups had heard of SI. Awareness did appear to vary according to geographical location, however, with farmers in the Upper Welland more likely than expected, and farmers in the Taw and Conway less likely than expected, to have heard of the term (confirmed as significant using a chi-squared test for independence ($\chi^2=21.35$, $p=0.046$), though with low confidence due to 33 per cent cells having an expected count of less than 5). Awareness also increased with farm size from 44 per cent of small farmers to 61 per cent of ultra-large farmers, and appeared to be higher among arable farmers (63 per cent of both cereal and general cropping farmers had heard of SI) than livestock farmers (44 per cent of both LFA and lowland grazing livestock farmers had heard of it), although these findings could not be confirmed as statistically significant due to the small numbers of farms within some categories.

Farmers were also asked to explain what they understood by the term SI, regardless of whether or not they had previously heard of it. Morris et al. (2016) analysed and grouped these stated understandings of SI into four themes, as follows:

- **Increased agricultural production while taking into account the environment (23 per cent).** E.g.:

  > Maintaining a high level of productivity while keeping or improving the environment. Without pollution or poisoning anything.
  >
  > (Very large cereal farm, Nafferton)

  > Increasing production without beating up the environment.
  >
  > (Large LFA Grazing Livestock farm, Taw)

- **Agricultural output and the current/future state of the farm business (18 per cent).** E.g.:

  > [SI is a] business term, to retain farming as a viable economic enterprise.
  >
  > (Very large dairy farm, Eden)

  > I've heard of the two words but not together like that. But intensification as in intensive farmer, sustainable as financially viable.
  >
  > (Ultra-large mixed farm, Avon)

- **Addressing the environmental impacts of agriculture only (7 per cent).** E.g.:

> Well like many farmers I suppose I realise that unless we treat the land with respect it's not going to be available to us for the future, so it's a matter really I suppose of reverting back in some degrees to what I would consider old-fashioned methods. And not reliant quite so much on chemicals and fertilisers.
>
> (Large cereal farm, Upper Welland)

> Going more green.
>
> (Large dairy farm, Eden)

- **Unable or unwilling to provide a definition of SI (51 per cent).** E.g.:

> Well, I can interpret it, but not really.
>
> (Small lowland Grazing Livestock farm, Taw)

> Not sure – increasing output but with sustainability, but not clear definition of what is being sustained.
>
> (Very large cereal farm, Upper Welland)

This analysis revealed that, while just over half of farmers had heard of the term SI – arguably already a relatively low proportion given the emphasis placed on SI in current agricultural policy and advice – even fewer understood it in a manner aligning with official definitions as relating to both agricultural production and the environment (theme 1). Again, these responses varied according to case study area, with farmers in the Wensum & Yare and Nafferton catchments appearing particularly well informed about SI as relating to both environmental and production objectives, and farmers in the Taw particularly unsure of its meaning (see Figure 5.4): These findings were confirmed as statistically significant ($\chi^2=43.18$, $p=0.001$). Similarly to our findings around awareness of the term, understanding of SI as relating to both production and environment goals increased with farm size from 13 per cent of small farms to 33 per cent of very large farms, with the slight exception of ultra-large farms (30 per cent), who were more likely than other farms to understand SI as relating to business/production goals only (33 per cent offered this type of definition compared to 18 per cent of all farms). Interpretation of SI as relating to both production and the environment was also higher among arable farmers than livestock farmers (38 per cent of general cropping farmers and 34 per cent of cereal farmers offered such definitions, compared to 20 per cent of LFA and 18 per cent of lowland grazing livestock farmers), though again these findings were unable to be confirmed as statistically significant due to small sample sizes within each category.

It is plausible that smaller farms are, by virtue of time pressures etc., less involved in farmer networks and have less access to social and formal learning opportunities than larger farms, and that the traditionally managed nature of many livestock units leaves these farmers less exposed to new policies, technologies and related SI practices. While this explanation is, to an extent, speculation on our part, it is one that is supported by wider SIP findings (based

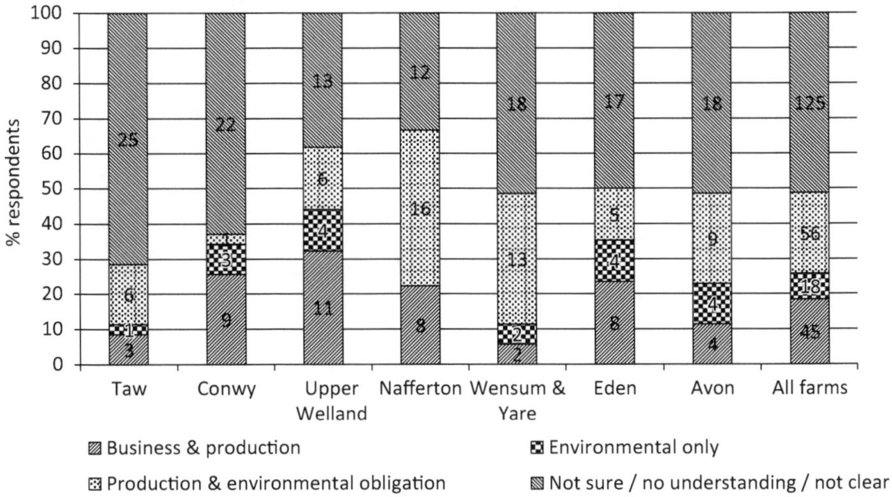

Figure 5.4 Farmer understanding of the term sustainable intensification

Source: Morris et al., 2016

on farmer discussion groups) (Fish, 2017) and Farm Business Survey (FBS) analysis (Wilson, 2017) showing that small farms are less likely to be involved in collaborative activities than larger operators. Some SIP participants also 'claimed that some sectors of farming were simply harder to reach, with the figure of the "isolated livestock" farmer offered as an individual who struggles to naturally collaborate and procure information in these [discussion group] settings' (Fish, 2017, p.40). Further evidence for this comes from a study by Dwyer et al. (2007), which suggests that;

> Overall, larger farm businesses, particularly arable ones, appear to create a certain amount of 'space' for their farmers to reflect and take time out to attend events and maintain social and business networks. By contrast, smaller livestock enterprises and most obviously, small to medium-sized dairy farms suffer particularly from a lack of time to do anything more than cope with the day-to-day business of running the farm. As a result, both business and community networking suffer.
>
> (Dwyer et al., 2007, p.36)

While reasons for differing levels of awareness and understanding between case study areas are difficult to determine, the apparent relationship with farm size and type might offer a part explanation. For instance, the Taw catchment – where levels of both SI awareness and understanding were lower than elsewhere – is characterised by relatively small, livestock-oriented farming which, given the discussion above, might account for some of the variance between this and other areas. That said, lower awareness of SI in the Taw catchment is nevertheless surprising given that Farming and Wildlife Advisory Group (FWAG) had been actively promoting the concept in this area immediately prior to the survey (Morris et al., 2016). In contrast, the livestock-dominated area of the Nafferton counters the tendency for lower awareness among this farm type, perhaps indicating particularly effective communication of SI in this area; though the larger size of farm units here in comparison to the Taw (and Conwy) may be implicated in this finding. Finally, it is interesting to note that the farming press and 'other media' – which in theory all farmers have access to – were found to be the most important sources of farmer awareness of SI (mentioned by 33 per cent and 20 per cent of all respondents respectively in response to an open question, 'Where have you heard of SI?').

When talking about awareness and understanding of SI, 8 per cent were critical of one or more aspects of the concept. It is perhaps surprising that higher levels of criticism were not expressed, but we should bear in mind that farmers were not specifically asked about their opinion on the subject, or to identify problematic aspects. The usefulness of the term, and achievability of the concept more generally, was challenged both by farmers who had offered their own definition of SI and those who had not. Examples of comments from those who had defined SI include:

> They want it to mean that we can produce as much food as possible with the least possible damage to the environment and land. Reduce inputs but increase output. It's a nonsense.
>
> (Very large mixed farmer, Nafferton)

> It's a buzzword, more glorified than it is.
>
> (Very large cereal farmer, Avon)

> As much output as possible without damaging the land. I am doing it anyway.
>
> (Medium cereal farmer, Nafferton)

Comments from respondents offering no clear definition of SI followed similar lines. For example;

> Another quango talking rubbish.
>
> (Large mixed farmer, Upper Welland)

> Two diametrically opposed ideas – Intensification is not sustainable. The word sustainable has been misunderstood and hijacked by the sound-bite brigade.
>
> (Small dairy farmer, Eden)

> Is intensive farming sustainable?
>
> (Large lowland grazing livestock farmer, Nafferton)

Farmer involvement in SI activities

Despite relatively low levels of awareness and understanding of SI, analysis of the practices that farmers said they were currently carrying out reveals reasonable levels of take-up (where applicable) for most activities; see Table 5.3.

Table 5.3 Involvement in SI activities (% respondents)*†

SI Activity (summary)	*Already carry out*	*Would consider introducing/ Increasing*	*Would not consider*	*Not applicable to farming system*
Grow crop varieties with increased tolerance to stresses such as drought, pests or disease	43	18	6	33
Reduce tillage to minimum or no till	44	14	9	33
Incorporate cover crops, green manures and other sources of organic matter to improve soil structure	44	19	11	25
Improve animal nutrition to optimise productivity (& quality) and reduce the environmental footprint of livestock systems	59	13	9	19
Reseed pasture for improved sward nutrient value and/or diversity	59	14	15	12
Predict disease and pest outbreaks using weather and satellite data	29	26	23	21
Adopt precision farming: using the latest technology (e.g. GPS) to target delivery of inputs (water, seeds, pesticides, fertilisers, livestock manures)	31	31	18	21
Monitor & control on-farm energy use	45	24	16	15
Improve the use of agriculturally marginal land for natural habitats to provide benefits such as soil improvement, pollution control or pollination, and allow wildlife to thrive	80	10	6	4
Provide training for farm staff on how to improve sustainability/ environmental performance	18	16	11	54

Notes: * Figures have been rounded up.
† Columns do not sum to 100% as respondents may be involved in more than one activity.

Source: Morris et al., 2016

For instance, the first five activities listed each had take-up rates of 59–73 per cent when excluding those farms where they were not applicable. 'Optimise marginal land for ecosystem services' had a particularly high take-up rate of 83 per cent (where applicable). While in one sense these figures are encouraging, the findings suggest a certain disjuncture between understanding and practice, which is problematic because if farmers are not fully buying into the concept of SI they are less likely to be committed to delivering it in the long term (Morris et al. 2016). It is also interesting to note that the activities requiring newer technologies (predicting disease and pest outbreaks using weather and satellite data, and precision farming) currently have the lowest levels of take-up (37 and 39 per cent where applicable respectively), along with 'train staff for improved sustainability' (39 per cent where applicable). We discuss some of the barriers constraining the use of SI activities below.

Barriers to SI activities

Farmers were asked an open question about whether there were any particular barriers that would prevent them from considering the SI practices listed. Their responses were subsequently coded and grouped into common themes: the ten barriers most commonly identified are shown in Table 5.4.

Table 5.4 Top 10 barriers to undertaking SI activities cited by farmers

Barrier	*No. comments*
Financial cost	99
Unconvinced of benefit (either environmental and/or agronomic) re a specific activity	79
Reduce tillage	*24*
Precision farming	*13*
Improve animal nutrition	*9*
Predict disease and pest outbreaks	*8*
Incorporate cover crops, green manures etc.	*8*
Reseed pasture	*6*
Train staff for improved sustainability	*3*
Grow tolerant crop varieties	*2*
Optimise marginal land for ecosystem services	*2*
General comment	*1*
Unsuitable for type of land or farming system	29
Activity restricted by legislation or scheme (e.g. AES, organic)	28
Lack of skills, knowledge and/or information	26
Lack of time	21
Already doing enough	11
Tenancy related barriers	11
Age and/or winding down the farm business	10
Concern re impact on agricultural productivity	10

Source: Morris et al., 2016

Unsurprisingly, the most commonly cited barrier listed was financial cost, with 41 per cent of farmers mentioning this. This was a concern across all farm types and sizes, although small farmers particularly referred to economies of scale and believed many activities to be unviable on their particular farm. Financial cost also appeared to be a particular barrier to adopting 'hi-tech' practices such as precision farming. The following sentiment was typical:

> The set-up costs [for precision farming] are huge for my farming size. We're already doing it in a rudimentary way – for instance, soil sampling in quarters of each field to target the use of P and K. I know people who have done it and the feedback from a lot is that they don't feel they've had the returns for the set-up and ongoing costs they've laid out.
>
> (Very large cereal farm, Avon)

Many farmers also remained unconvinced about the benefit of one or more of the SI activities, either for the environment and/or for the success of their business (and were perhaps unwilling to invest in relevant equipment for this reason). This scepticism was present to some extent for most of the activities listed ('monitor and control on-farm energy use' was the only activity entirely unchallenged), but particularly emerged in relation to 'reduce tillage to minimum or no till', which was questioned by 24 farmers (10 per cent). These farmers argued that reducing tillage was not suitable for their type of land and/or resulted in problems such as weeds and pests. For instance:

> We're on heavy clay loam so I don't think we'd get the drainage we need. It wouldn't suit the land quality – it's a different thing 2 miles away which is red sandstone – a few of them have got min-till but most of them still plough. The problem we've got with long-term lays is that the ground becomes compacted on the top and by ploughing it we can break that up. To plough it up and reseed it has made a tremendous difference – it's a lot drier to walk over.
>
> (Large dairy farm, Taw)

> We try to min-till but it's not quite working to plan. It's quite difficult because we're using maximum rates of spray on the clover but it takes a lot of killing off. So it looks as though will have to plough the field up because don't want the clover to become too dominant.
>
> (Large mixed farm, Avon)

'Adopt precision farming' and 'improve animal nutrition to optimise productivity (and quality) and reduce the environmental footprint of livestock systems' were also particularly challenged as not being beneficial/necessary (mentioned by 13, 9 and 8 farmers respectively), either because of the size of the farm and/or because the farmers were content with their current practices and did not see a need to improve them. This was also the case for 'predict disease and pest outbreaks using weather and satellite data', although some farmers also felt that this type of technology was not yet advanced enough to be sufficiently accurate. Inappropriateness to the individual farm – particularly on the basis of small size or extensive farming system – was also mentioned more generally (i.e. not in relation to a specific activity listed) by 29 farmers (12 per cent) as a reason for not adopting (or increasing) SI activities.

Farmers' concerns about the efficacy and suitability of SI practices are entirely justified in many cases, and their knowledge and experience of farming their land in the most appropriate manner should not be underestimated. There is, however, clearly scope for improving and demonstrating SI-related technologies. Early adopters are key in this process – as one farmer said, 'before we commit a lot of resources or big investment we tend to look for other farmers that are doing it' – but demonstration farms and solid evidence proving the benefits of practices are also important. For instance, several farmers stated they were open to considering incorporating cover crops and green manures, but were cautious about doing so without further substantiation of its value:

> There needs to be more trial work on proving their cost-effectiveness. Our neighbour is doing it and our agronomist also works for him, so watch this space. We've got to see the economic advantage of doing so. But we all know that our organic matter levels in the soil are dropping and anything we can do to stem that or maintain it is a good thing to do.
>
> (Ultra-large general cropping farm, Wensum and Yare)

> I'm unsure of the evidence – I think the practicalities are yet to be fully demonstrated. But if they were we'd certainly consider them.
>
> (Very large cereal farm, Upper Welland)

> There is a change in legislation and quite a bit of research around that – is it appropriate for European farming? Watch this space. If it's deemed appropriate it will get done.
>
> (Very large general cropping farm, Wensum and Yare)

In relation to this issue of evidence, it is also pertinent to note that 26 farmers (11 per cent) mentioned a lack of skills, knowledge and/or information as barriers to adopting more SI practices. In some cases, this was about not having the personal skills or confidence to employ what they perceived as hi-tech, 'modern' agronomic practices:

> [re: barriers to precision farming] Me [laughs], because I didn't go to university. I have modern machinery, which is ready for GPS, but I don't use it at the moment, I don't have the confidence with the technology … My wife is trying to persuade me to use it, so I might do in the future. I expect future generations are likely to use it.
>
> (Very large LFA grazing livestock farm, Taw)

> [re: barriers to precision farming] One of the biggest problems is our age – we're not technology whizz-kids. We have got a young lad on the farm but he's not – well, he's better than us but he's not [an expert] ... We're all over 60 so it's not something that's, to be fair it is in its infancy and we've never been at the forefront of technology.
>
> (Very large cereal farm, Wensum and Yare)

For others, however, it was simply a case of 'needing to find out more'.

> I don't know a lot about precision farming, I'm relying on the contractors and agronomist really. I suppose we're doing it in a sense. Cost comes into it, it has to be cost efficient, but it's generally just about finding out more about these things. I'm looking for advice and information, so I'm open to considering all these things.
>
> (Medium lowland grazing livestock farm, Hampshire Avon)

> The trouble is the advice changes all the time. And I'm not sure how it would affect our land, our production, our costs.
>
> (Medium lowland grazing livestock farm, Hampshire Avon)

> There's nothing to stop me really, it's just about finding out more about them.
>
> (Large general cropping farm, Wensum and Yare)

Most of the other barriers listed in Table 5.4 perhaps speak for themselves, but it is worth considering for a moment that 28 farmers (12 per cent) stated that adoption of one or more of the SI practices listed was constrained due to a restriction imposed under legislation or a scheme they were involved with. Of particular note, eleven of these comments referred to an agri-environment agreement, five to farming organically and six to regulations concerning genetic modification (GM) and/or pesticide usage.[2] The comments concerning AES highlighted potential contradictions between (and within) some of the activities identified as 'SI' and some of the practices advocated under agri-environment policies. These included complications regarding low-inputs and min-till because of the inability to spray weeds (this was also specifically cited as an issue by two of the organic farmers), the inability to use cover crops due to being 'locked in' to over-winter stubbles and, most notably, reseeding being specifically prohibited under the schemes. As one farmer explained;

> Reseeding pasture is against our HLS agreement – it's permanent grassland – so you've got a conflict of interest there. We did reseed a meadow in 1976 but within a year the indigenous grasses were back. So I thinks it's a waste of time anyway, it's better to rotate.
>
> (Very large general cropping farm, Wensum and Yare)

The six farmers who mentioned wider regulations around GMs and chemicals cited these as particular barriers to; the development of crop varieties with increased tolerance to stresses; the use of no- or min-till (due to not being able to control weeds sufficiently); and/or the general ability to increase productivity to meet SI objectives. These farmers strongly argued for liberalisation of the industry in this sense. For instance;

> You can only do that [grow tolerant crop varieties] within the breeding system that we're allowed to adopt. GM could help improve those traits much, much quicker. We have to move in that direction because pesticide restrictions are going in such a way that – led by the blundering idiot of the EU – and ultimately the regulations on pesticides, whether or not they're safe to use, are closing them down or taking them off the marketplace. There's a massive issue there.
>
> (Ultra-large general cropping farm, Wensum and Yare)

> We need all the technology we can being used and the EU's terrible at it – it needs to free up quite a bit. Compared to New Zealand, where there is an enabling policy environment, here it is a disabling one. We're not growing as much as we could do – we're politically restrained. I mean the GM arguments – it's just pants really isn't it. There's no give and take, which I think is very sad actually.
>
> (Very large general cropping farm, Wensum and Yare)

> The inability to use GM due to European restrictions is a major barrier to SI, as that could really increase resistance to drought and disease. We have a huge problem with the loss of actives we're allowed to use. When I started farming there were over 1,000 actives you could use in crop protection, but now that's down to 250. Europe is imposing bans on a range of products that are available elsewhere in the world and that is a huge barrier to us now. Politically it's becoming one of the most important issues. We're competing on the world market with countries who still use the products we ban. There needs to be a level playing field. For example, neo-nic[otinoid] seed dressings – the most widely used one is on sale in Australia and they claim to have the healthiest bees in the world. We need to be able to use more, otherwise we have to just spray everything. And with climate change and more extreme events the issue becomes even more important.
>
> (Ultra-large cereal farm, Hampshire Avon)

As with so much that is Brexit related, the issue of future policy towards GM is complex. In the short term (i.e. during the transitional phase) it is unlikely that anything will change. In the longer term, the UK's position towards GM could be influenced by trade deals but it appears that the government may review the policy. In response to a written parliamentary question, George Eustice (Minister of State for Agriculture, Fisheries and Food) reported that 'As part of the preparations for EU exit, the Government is considering possible future arrangements for the regulation of genetically modified organisms. The Government's general view remains that policy and regulation in this area should be science-based and proportionate' (Eustice, 2016).

Conclusions

In policy and research terms SI received significant exposure in the years after 2010. But there are continuing ambiguities and uncertainties regarding the conception. We have seen this from within the farming community in this chapter. But research has shown British public ambivalence too with public opinion more concerned with sustainable consumption and greater equity than with production-based technological SI solutions (Barnes et al., 2016). Nor is there unanimity and clearly defined terminology in the expert community. Petersen and Snapp (2015) show that the term SI is variably interpreted by agricultural experts (in the US) and most do not see it as offering anything new to current practices. They suggest that it is not clearly understood by agronomists, who tend to focus only on production and technological aspects. Some have argued that SI and food security can be achieved through technology as long as adequate investments are made (Flavell, 2010) but this is to ignore the wider context of sustainability, especially the social question. If SI remains ambiguous and experts are sceptical, is it a useful paradigm? Might a more holistic approach be achieved by the notion of ecological intensification instead? These are hard questions. As we have seen in this chapter, there have been many attempts already to broaden and deepen the scope of SI and it is widely accepted that SI's full-scale potential cannot be achieved without overcoming significant economic, social and policy related barriers (Barnes, 2016). And it is in this context that we need to consider the rather modest farm-level evidence reported in this chapter and in other studies such as that of Firbank et al. (2013), who provide evidence of SI narrowly defined (as increased food production while reducing pollution and enhancing biodiversity) and motivated by improving farm profitability through increased production and efficiency and accessing agri-environment funding, rather than in response to the SI agenda per se.

A final concluding observation is that the 'social' aspects of 'sustainability' in SI remain ill-developed, whether it be the wider social context of connections between farmers and the public as seen through the lens of human health and well-being, or the well-being of farmers and farm workers themselves. The SI Research Network set up in the aftermath of SIP is funded by the UK Biotechnology and Biological Sciences Research Council and Natural Environment Research Council but *not* the Economic and Social Research Council. And yet within policy circles it is widely held that the main challenge facing British agriculture in a Brexit context is not fundamentally a lack of scientific understanding of technological, ecological and agronomic desiderata, but the social and economic context. Farmers are not necessarily ready adopters of SI innovations in a Brexit world because socially and culturally, and indeed in actuality because policy has not yet changed, all too often they inhabit a world of Basic Payments and social isolation.

Notes

1 The various SIP reports are available through www.siplatform.org.uk/outputs.
2 Of the remaining six comments, two referred to red tape generally, one to RB209, one to National Park restrictions, one to archaeological status, and one to an unspecified restriction concerning drainage.

References

Anthony, S., Boatman, N., Cosby, J., Crowe, A., Emmett, B., Henrys, P., Hodge, I., Lee, D., Midmer, A. & Thomas, A. (2016). *Landscape Typology – A Framework for Prioritisation of Sustainable Intensification Strategies*. Report for Defra project LM0302 Sustainable Intensification Research Platform Project 2: Opportunities and Risks for Farming and the Environment at Landscape Scales.

Barnes, A. P. (2016). Can't get there from here: Attainable distance, sustainable intensification and full-scale technical potential. *Regional Environmental Change*, 16(8), 2269–2278.

Barnes, A. P., Lucas, A. & Maio, G. (2016). Quantifying ambivalence towards sustainable intensification: an exploration of the UK public's values. *Food Security*, 8(3), 609–619.

Brisson, N., Gate, P., Gouache, D., Charmet, G., Oury, F.-X. & Huard, F. (2010). Why are wheat yields stagnating in Europe? A comprehensive data analysis for France. *Field Crops Research*, 119(1), 201–212.

Buckwell, A., Heissenhuber, A. & Blum, W. (2014). *The Sustainable Intensification of European Agriculture: A review sponsored by the RISE Foundation*. Brussels: Rural Investment Support for Europe (RISE).

Clay, D., Reardon, T. & Kangasniemi, J. (1998). Sustainable intensification in the highland tropics: Rwandan farmers' investments in land conservation and soil fertility. *Economic Development and Cultural Change*, 46(2), 351–377.

Cook, S., Silici, L., Adolph, B. & Walker, S. (2015). *Sustainable Intensification Revisited*. London: IEED Issues Paper.

Defra (2018). *Structure of the agricultural industry in England and the UK at June*. [Online] Available at: www.gov.uk/government/statistical-data-sets/structure-of-the-agricultural-industry-in-england-and-the-uk-at-june [accessed 12/03/18].

Defra & Rural Business Research (2018). *Farm Business Survey (FBS) data builder*. [Online] Available at: www.farmbusinesssurvey.co.uk/DataBuilder/ [accessed 05/02/18].

Dwyer, J., Mills, J., Ingram, J., Taylor, J., Burton, R., Blackstock, K., Slee, B., Brown, K., Schwartz, G., Matthews, K. & Dilley, R. (2007). *Understanding and influencing positive behaviour change in farmers and land managers – a project for Defra*. Gloucester: CCRI, Macaulay Institute.

Eustice, G. (2016). Genetically modified organisms: written question – 48641. [Online] Available at: www.parliament.uk/business/publications/written-questions-answers-statements/written-question/Commons/2016-10-13/48641 [accessed 27/06/18].

Firbank, L. G., Elliott, J., Drake, B., Cao, Y. & Gooday, R. (2013). Evidence of sustainable intensification among British farms. *Agriculture, Ecosystems & Environment*, 173(July), 58–65.

Fischer, R. A. (2015). Definitions and determination of crop yield, yield gaps, and of rates of change. *Field Crops Research*, 182(October), 9–18.

Fischer, T., Byerlee, D. & Edmeades, G. (2014). *Crop Yields and Global Food Security: Will yield increase continue to feed the world?* ACIAR Monograph No. 158. Canberra: Australian Center for International Agricultural Research (ACIAR).

Fish, R. (2017). *Farmer Discussion Groups – Key findings*. Report for Defra project LM0302 Sustainable Intensification Research Platform Project 2: Opportunities and Risks for Farming and the Environment at Landscape Scales.

Fish, R., Winter, M. & Lobley, M. (2014). Sustainable intensification and ecosystem services: New directions in agricultural governance. *Policy Sciences*, 47(1), 51–67.

Flavell, R. (2010). Knowledge and technologies for sustainable intensification of food production. *New Biotechnology*, 27(5), 505–516.

Gunton, R. M., Firbank, L. G., Inman, A. & Winter, D. M. (2016). How scalable is sustainable intensification. *Nature Plants*, 2(5), 16065.

Hazell, P. B. R. (1995). *Managing agricultural intensification*. 2020 policy brief 11. Washington D.C.: International Food Policy Research Institute

Henrys, P., Anthony, S., Boatman, N., Brown, M., Cooper, J., Cosby, J., Crowe, A., Emmett, B., Johnson, C., Lee, D., Midmer, A., Thomas, A. & Watkins, J. (2016). *Dynamic Typology Tool – Initial Development and Future Roadmap*. Report for Defra project LM0302 Sustainable Intensification Research Platform Project 2: Opportunities and Risks for Farming and the Environment at Landscape Scales.

Knight, S., Kightley, S., Bingham, I., Hoad, S., Lang, B., Philpott, H., Stobart, R., Thomas, J., Barnes, A. & Ball, B. (2012). *Project Report No. 502: Desk study to evaluate contributory causes of the current 'yield plateau' in wheat and oilseed rape*. Kenilworth: AHDB.

Kuyper, T. W. & Struik, P. C. (2014). Epilogue: Global food security, rhetoric, and the sustainable intensification debate. *Current Opinion in Environmental Sustainability*, 8, 71–79.

Loos, J., Abson David, J., Chappell, M. J., Hanspach, J., Mikulcak, F., Tichit, M. & Fischer, J. (2014). Putting meaning back into "sustainable intensification". *Frontiers in Ecology and the Environment*, 12(6), 356–361.

Mackay, I., Horwell, A., Garner, J., White, J., McKee, J. & Philpott, H. (2011). Reanalyses of the historical series of UK variety trials to quantify the contributions of genetic and environmental factors to trends and variability in yield over time. *Theoretical and Applied Genetics*, 122(1), 225–238.

McCalla, A. (1995). Towards a strategic vision for the rural/agricultural/natural resource sector activities of the World Bank. *World Bank 15th Annual Agricultural Symposium. 4 January, Washington D.C.*

Morris, C., Jarratt, S., Lobley, M. & Wheeler, R. (2016). *Baseline Farm Survey – Final Report*. Report for Defra project LM0302 Sustainable Intensification Research Platform Project 2: Opportunities and Risks for Farming and the Environment at Landscape Scales.

Mueller, N. D., Gerber, J. S., Johnston, M., Ray, D. K., Ramankutty, N. & Foley, J. A. (2012). Closing yield gaps through nutrient and water management. *Nature*, 490(7419), 254.

NFU (2015). *UK agricultural productivity fails to keep pace with global trends* [Online]. Available at: www.nfuonline.com/cross-sector/farm-business/economic-intelligence/economic-intelligence-news/uk-agricultural-productivity-fails-to-keep-pace-with-global-trends/ [accessed 16/10/17].

Petersen, B. & Snapp, S. (2015). What is sustainable intensification? Views from experts. *Land Use Policy*, 46, 1–10.

Pretty, J. N. (1995). Participatory learning for sustainable agriculture. *World Development*, 23(8), 1247–1263.

Reardon, T., Crawford, E., Kelly, V. & Diagana, B. (1995). *Promoting Farm Investment for Sustainable Intensification of African Agriculture*. Food Security International Development Papers 54053. Michigan State University: Department of Agricultural, Food and Resource Economics.

Reardon, T., Kelly, V., Crawford, E., Diagana, B., Dioné, J., Savadogo, K. & Boughton, D. (1997). Promoting sustainable intensification and productivity growth in Sahel agriculture after macroeconomic policy reform. *Food Policy*, 22(4), 317–327.

Rockström, J., Williams, J., Daily, G., Noble, A., Matthews, N., Gordon, L., Wetterstrand, H., DeClerck, F., Shah, M., Steduto, P., de Fraiture, C., Hatibu, N., Unver, O., Bird, J., Sibanda, L. & Smith, J. (2017). Sustainable intensification of agriculture for human prosperity and global sustainability. *Ambio*, 46(1), 4–17.

Sylvester-Bradley, R., Kindred, D. R., Blake, J., Dyer, C. J. & Sinclair, A. H. (2008). *Optimising Fertiliser Nitrogen for Modern Wheat and Barley Crops*. HGCA Final Project Report 438. Kenilworth: HGCA/AHDB.

Wilson, P. (2017). *Analysis of Farm Business Survey 2011–12 Business Management Practices*. Report for Defra project LM0302 Sustainable Intensification Research Platform Project 2: Opportunities and Risks for Farming and the Environment at Landscape Scales.

Winter, M., Oliver, D. M., Fish, R., Heathwaite, A. L., Chadwick, D. & Hodgson, C. (2011). Catchments, sub-catchments and private spaces: Scale and process in managing microbial pollution from source to sea. *Environmental Science & Policy*, 14(3), 315–326.

YEN (2018). *Cereal yield potential* [Online]. Available at: www.yen.adas.co.uk/About/CEREALYEN.aspx [accessed 09/04/18].

Zayed, Y. (2016). *Agriculture: Historical statistics*. Briefing Paper No. 03339. London: House of Commons Library.

6 The small farm question

Introduction

Having considered the relatively new issue of sustainable intensification, we now turn to an enduring issue of debate and concern; the role and fate of small farms. Advocates of small family farms regard their contribution as distinct and worthy of support. In this chapter, we draw on work conducted for the Prince's Countryside Fund on the future of the small family farm (Winter and Lobley, 2016) to examine the extent to which these claims for the virtues of small or family farming can be justified and, if vindicated, what might be done to improve viability and resilience in this part of the farming industry. Our focus on small farms here is justified on the basis that they are significant components of the UK agricultural industry (numerically, if nothing else: see Chapter 3) and that they face particularly difficult challenges at the current time.

Does size matter?

The literature clearly shows that the term 'small farm' can have different meanings depending on the research or policy context (see Bonanno, 1987; Lobley, 1997; Pritchard et al., 2007). Many definitions of farm size are used across the policy and research spheres, including measures based on land area, Standard Gross Margins (SGM), Standard Output (SO) and standard labour requirements (SLR). All farm size measures are imperfect to some extent, but crucial for present purposes is that, whichever method is applied, it is the relative size of farms that matters. Thus attention is focused in this chapter on the smallest size groups in terms of both land area (the geographical size of the holding) and aggregate SLRs (a measure based on standard figures for the labour requirements associated with different livestock and crop types), as these are the methods most commonly encountered in official sources such as Defra. Through concentrating our attention on small farms, there is also an implied focus on *family* farms. The extent of the decline of use of hired labour in agriculture is such that many farm businesses are family farms in terms of being family-owned businesses worked both managerially and operationally by family members. The vast majority of small farms are family farms in these terms, though not all family farms are small.

The merits or otherwise of *small* family farms have not been as keenly debated in Britain in recent years as was once the case, and this is reflected in a relatively modest recent literature on the influence of farm size and the positive or negative aspects of small farming. This is in marked contrast to lively debate on small-scale agriculture in many other parts of the world (Akram-Lodhi and Kay, 2009; Brookfield and Parsons, 2007). Whereas in many European countries small or peasant farmers have long been portrayed as the backbone of rural society and custodians of the land (Fennell, 1979; Hoggart et al., 1995), and in the US there is a lively debate on the contribution of small farms to rural economy and society (Berry, 2002; Hayes-Conroy, 2007), in the UK, or more especially England, a positive relationship between small farm size and a sustainable countryside is not so easily claimed. Within mainstream contemporary agriculture, size or scale of activity is usually now seen as less important when measuring farm characteristics and level of performance, whether economic or environmental, than a whole set of other variables (for example, farmer attitudes and type of enterprise).

Today, it is probably fair to say that in many quarters there is a carefully cultivated neutrality on the question of size; a consequence of two factors. First, the average farm size in Britain is larger than in many other European countries. This is linked to the estate system and the tri-partite (landlord-tenant-worker) model, so different to the peasant proprietorship systems of much of Europe (Cleary, 1989; Kopsidis, 2012; Van der Ploeg, 2003). Second, size neutrality perhaps reflects the political success of the National Farmers' Union (NFU) over many years in speaking for the industry as a whole. It is no coincidence that the NFU was founded more than a century ago amid considerable political debate over land reform and the role of the small farm in society (Brown, 2000; Cox et al., 1991; Cragoe and Readman, 2010; Flynn et al., 1996). Since its inception, the NFU has, on the whole, successfully contained the tensions between horn and corn and between small and large farmers.

But there are two qualifications to this narrative of 'size not mattering'. First, it is rather an English story. In Wales, the concern for the small or family farm was so strong in the 1950s that the Farmers' Union of Wales was born, in opposition to the NFU, explicitly to speak for the interests of the family farm (Murdoch, 1995; Winter, 1996). More recently it has been argued that 'the family farm defines the character of Welsh rural society, and its sense of identity. The numbers directly and indirectly involved in farming make a crucial contribution towards sustaining rural communities' (National Assembly for Wales, 2001, p.4). And in Scotland, in part because of the politics of crofting – subject to a recent major official investigation (Committee of Inquiry on Crofting, 2008) – there is currently a Small Farms Grant Scheme. In Northern Ireland, where farms have historically been smaller than elsewhere in the UK and where there is a different land tenure system, small farms continue to exert an influence both culturally and politically (Gosling, 2015). So policy and cultural resonance related to small farms varies significantly across the UK. Second, a strong positive discourse surrounding small farms has continued among some of those resistant to mainstream conventional agriculture. Thus, writing in *The Guardian*, the former head of the Soil Association characterises small farmers as the 'backbone of the rural economy':

> By their very existence, they play a crucial role in maintaining our countryside. They are the stewards of our landscapes, field boundaries and hedgerows, the guardians of the fertility of the soils, the pastures, biodiversity, and the ancient green lanes of herding the cattle in to be milked.
>
> (Holden, 2015 [online])

Others lament the decline of the small farm in a global context. Chris Smaje, who runs a website called 'Small Farm Future', writes:

> From the brief high-water mark of pro-peasant populism in the earlier part of the twentieth century, the possibility of founding self-reliant national prosperities upon independent small proprietors has slowly been eroded through land grabs, global trade agreements and agrarian policies favouring capital intensive staple commodity production over local self-provision, regardless of the consequences for small-scale farmers.
>
> (Smaje, 2015, p.7)

The close association between advocacy of small-scale farming and advocacy of radical organic alternatives to conventional agricultural systems (see Smaje, 2014; Tudge, 2007) often serves, in fact, to keep the size issue on the margins of mainstream debate. This is unfortunate in our view as there is real scope for positive interaction between alternative visions for agriculture and the concern at the challenges facing more conventional mainstream family farms. Two organisations that seek to make the case for small farmers, without any necessary link to organic systems, are the Small Farms Association and the Family Farmers' Association. The Small Farmers' Association was formed in 1997 by farmers frustrated by the reluctance of the major farming unions to recognise the concerns, and support the needs, of the small farmer in their policies and activities. The Family Farmers' Association dates back to 1979. As their website states, their purpose is to:

> promote family farming – the Family Farmers' Association has been fighting for the survival of civilised farming on family farms since 1979. Family farmers produce significant quantities of high quality food, while caring for the countryside. They enrich rural communities, because family farming involves a lot of country people. They are an endangered species.

That all is not well with small or medium-sized family farming is well documented in a recent report for the Prince's Countryside Fund, which reveals the gravity of declining profits and the financial pressures on farmers (Andersons, 2016), some of whom are already beset by other challenges such as bovine tuberculosis (bTB). Agricultural commodity prices are cyclical and highly dependent on global markets, but small farms face problems that go well beyond the ups and downs of commodity prices. To put it another way, if there were to be a sudden price spike many small farmers would have a short-term benefit, but this would not alter some of the longer-term problems they face.

Continuity and change

The structural changes that have occurred in British agriculture over the last century (as presented in Chapter 3) have clearly had a significant impact on what most people think of as the traditional small family farm. Most significant, perhaps, is the declining numbers of small farms amid financial adversity and the wider amalgamation and intensification of farm units. Yet, as already noted, small farms continue to make up a significant proportion of holdings across the UK. As Brookfield and Parsons (2007, p.1) comment,

> the total disappearance of the family farm has been confidently predicted for almost a century and a half, and is still predicted today. While a great number have not survived into the twenty-first century, the fact that so many have done so, and in so many different lands, is remarkable.

This 'remarkable' persistence of small farms is in part due to the personal resilience, endurance and determination of farming families to survive, but it is also increasingly reliant on their decisions to implement adaptive changes within the business and household unit. In their review of work on the survival of family farming and restructuring in British agriculture Lobley and Potter (2004, p.499) state that:

> A chief conclusion from this work is that despite the numerical stability of family farms as institutional units, the nature of farm households and the pattern of land holding is undergoing significant change, with farm families becoming both more pluriactive on the one hand and increasingly subsumed to external capital influences on the other. At the same time, the connections between occupancy of holdings and the management of land are becoming ever more complex and differentiated in space, with an ever greater diversity of ways in which it is possible to be 'a farmer'.

Based on a survey of 255 farmers in six areas of England,[1] Lobley and Potter (2004, p.502) go on to describe

> the extent to which many farming families are long established in their locales, finding that almost one third of respondents were from families that had been farming in the same area for more than a century. In total, 84% of the sample operated established family farms (defined as at least the second generation of the family farming the same farm or in the immediate area of the first family farm). Just 8% were new entrants in the sense that they were the first generation of their family to farm and/or had not previously farmed or occupied a farm elsewhere.

These findings are reiterated by those of the 2016 SW Farm Survey, which point to the longevity of farming families, alongside considerable farm business change, in recent years. On average, farming families in the survey had been farming either the same farm or in the general vicinity for 105 years and (as in Lobley and Potter's research) only 8 per cent were new entrants.

Small farmers face very real economic challenges at the current time and, indeed, this has characterised their position for many years. Broadly speaking, farmers face two choices in order to cope with declining economic fortunes: either to focus on a farming solution (the 'traditional' way) or to redeploy resources away from agricultural production (the 'new' pluri-active way). In reality, it may be a combination of the two or farmers may vacillate between the two courses of action with periods of off-farm work generating income interspersed with a focus on the farm. There are, of course, two other options open to farmers. First, they may cease farming, either entirely through selling up the farm or by letting their land. Or, second, they might tighten the belt and continue business as usual. Most of the agricultural restructuring identified by Lobley and Potter (2004) took the form of 'traditional restructuring', such as farm expansion to spread fixed costs, significant enterprise change such as a switch from dairy to beef production, and reductions in the use of hired labour. Large farms were found to be those most likely to be growing in size (a pattern later confirmed by Lobley and Butler, 2010), while 'for many small and medium-sized family farms the picture is one of adaptation and retrenchment rather than determined disengagement from agriculture' (Lobley and Potter, 2004, p.503). Small farms were, however, also disproportionately likely to be found in a group of 'capital consumers' which, although accounting for only 4 per cent of the sample

> brings together all those who have been actively withdrawing assets from farming: 63% have reduced the size of their land holding and 33% have sold non-land assets. Typically, small farms are operated by elderly or retired farmers (78% are over 55) and many of these are … 'retirement holdings' occupied by individuals at the end of their farming careers, often uncertain of succession but unable or unwilling to give up farming entirely.
>
> (Lobley and Potter, 2004, p.506)

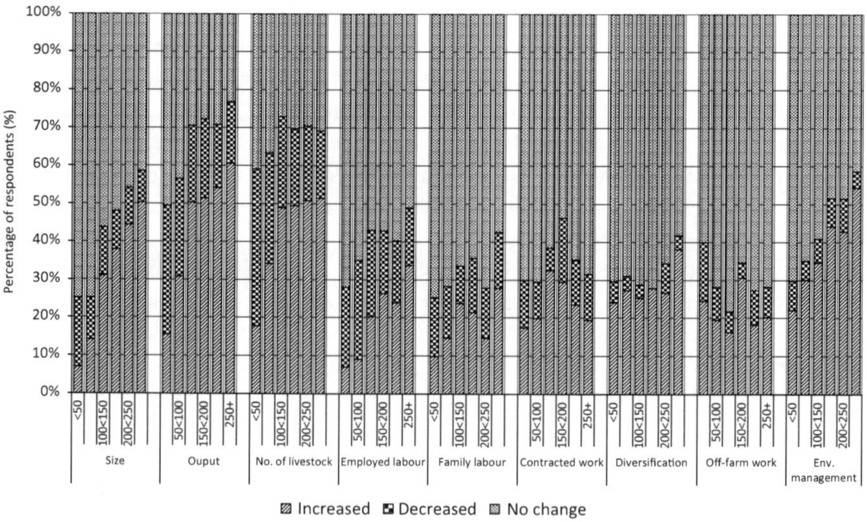

Figure 6.1 Changes made in the farm business since 2010, by farm size

Notes: Chi-squared values are as follows: Size χ^2 = 169.38, p<.001; Output χ^2 = 145.01, p<.001; No. livestock χ^2 = 103.91, p<.001; Employed labour χ^2 = 123.82, p<.001; Family labour χ^2 = 64.40, p<.001; Contracted work χ^2 = 38.05, p=.001; Diversification χ^2 = 30.34, p=.011; Off-farm work χ^2 = 35.29, p=.002; Env. management χ^2 = 83.63, p<.

Source: 2016 SW Farm Survey

Similar evidence of a tendency to consolidate or withdraw from farming among small farms emerged from the 2016 SW Farm Survey. Disaggregation of the data on changes made within the farm business over the past five years (discussed in Chapter 3) suggests that a group of the smallest farms (<50 ha) have been disengaging by reducing the area they farm and/or reducing livestock numbers and output, while expansion of both land area and output is clearly associated with larger farms (Figure 6.1). On the other hand, some small farms have expanded their farming activities and a significant proportion (almost a quarter in each case) appear to have been supplementing their agricultural income through diversification, off-farm work and/or increasing environmental management.

Looking to the future, evidence from the 2016 SW Farm Survey (Figure 6.2) suggests that overall, small farm size is significantly associated with a lower incidence of planned change, although it is notable that the operators of 20 per cent of the smallest farms are planning to increase diversification, suggesting that, for some, survival will continue to be predicated on the health of the wider economy. On the other hand, Table 6.1 indicates that for a significant proportion of operators of small farms, the near future will see them retiring or otherwise leaving agriculture. It is entirely appropriate in many instances that this should be so. Life decisions have to be made and

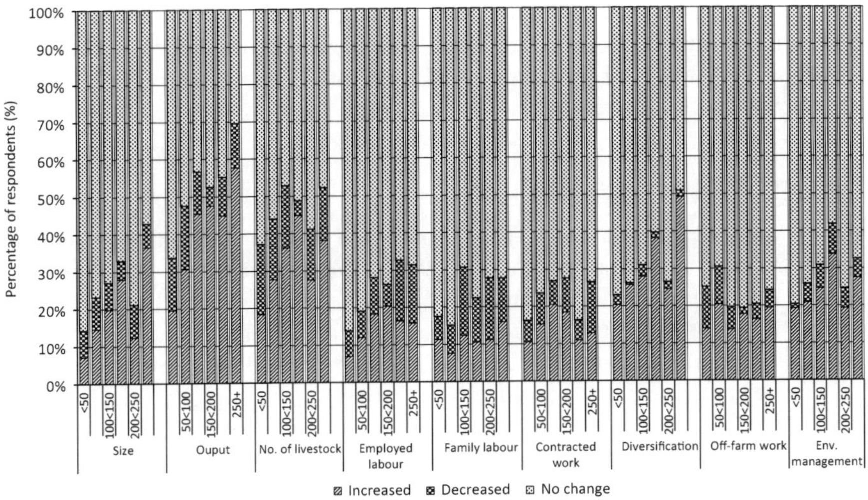

Figure 6.2 Planned changes in the farm business over the next five years, by farm size

Notes: Chi-squared values are as follows: Size χ^2 = 64.92, p<.001; Output χ^2 = 83.56, p<.001; No. livestock χ^2 = 40.10, p<.001; Employed labour χ^2 = 62.59, p<.001; Family labour χ^2 = 48.53, p<.001; Contracted work χ^2 = 28.26, p= .020; Diversification χ^2 = 61.75, p<.001; Off-farm work χ^2 = 24.28, p=.061; Env. management χ^2 =39.78, p<.001.

Source: 2016 SW Farm Survey

Table 6.1 Plans to retire or leave farming in the next five years (%)

	Farm size (ha)						*All farms*
	<50	*50<100*	*100 <150*	*150 <200*	*200 <250*	*>250*	
Plans to retire in the next 5 years	55.3	45.0	38.6	30.7	37.5	24.1	42.3
No plans to retire in the next 5 years	44.7	55.0	61.4	69.3	62.5	75.9	57.7

Note: χ^2=*49.48, p<.001.*

Source: 2016 SW Farm Survey

people retire from work. What is of concern is whether the ranks of these small farms can be replenished by active and economically vibrant new small farms or whether, as seems more likely unless current trends are modified or reversed, their land and property is taken up by a combination of expanding large farms and residential lifestyle purchasers.

The contribution of small farms

So why are small farms deemed by some to be so important and worthy of special attention and support? Here, we examine arguments and evidence relating to the contribution of small farms to the economy, rural communities and the rural environment. Importantly, we recognise that 'the case for small farms' is often built on assumptions of what might replace them, so we also consider whether, and how, the decline of small farms might impact on food production, rural communities and environmental management.

The economic contribution

It has been argued that 'the approximate measure of rural community well-being is and should still be employment' (Midmore and Dirks, 2003, p.3). This is because employment or, more precisely, *paid* employment is the most important means of achieving other ends. In terms of direct employment, evidence from the 2016 SW Farm Survey indicates that the 1,070 respondents supplying employment data employed 3,164.75 full-time equivalents (FTEs) (including family labour and those working in diversified enterprises). On average, small farms of less than 50 hectares employed just over 2 FTEs, compared to the largest farms employing 5.25 FTEs (see Table 6.2). However, significantly, it is smaller farms that employ more labour per unit area. Table 6.3 indicates that mean and median employment per 100 hectares is greater on smaller farms (<50 ha and <100 ha) compared to larger farms.

If this employment benefit is noticeable on mainstream farms, it is even more apparent on smaller less conventional farms. For instance, evidence from the Landworkers' Alliance based on a survey of 70 farmers operating farms of 20 hectares and less points to 'the labour intensive nature of farms in the survey, especially those with a horticultural element, means that more people are employed per hectare than is typical for most traditional farming

Table 6.2 Mean FTE, by farm size

	Farm size (ha)						*All farms*
	<50	*50<100*	*100<150*	*150<200*	*200<250*	*>250*	
Mean	2.15	2.09	2.95	3.05	4.85	5.25	2.96

Source: 2016 SW Farm Survey

Table 6.3 Mean and median FTE per 100 ha, by farm size (with outliers removed*)

	Farm size (ha)						*All farms*
	<50	*50<100*	*100<150*	*150<200*	*200<250*	*>250*	
Mean	7.72	2.95	2.51	1.78	2.27	1.23	3.64
Median	4.95	2.47	1.87	1.61	1.48	1.06	2.22

Note: *Ten cases have been removed as these were small business with a very large number of employees which were distorting the results. With the outliers included the mean FTEs per 100 ha for small farms (< 50 ha) was 12.49.

Source: 2016 SW Farm Survey

activities' (Redman, 2015, p.188). While family labour is the main input, holdings as small as 1 hectare were providing a livelihood for up to three FTEs.

It is also worth noting that, while more difficult to quantify, the contribution to employment from small farms is not confined to direct on-farm employment. Workshop participants in Winter and Lobley's (2016) research were keen to point out that the presence of small farms also supports the existence of associated shops and services in the wider rural economy, such as agricultural retailers, machinery suppliers and larger contracting firms. Furthermore, the many small farmers who have diversified provide a network of services to the community; for example, as contractors or repairers of machinery.

The other main economic/agricultural contribution of small farms is to agricultural output itself. Here, as with much of the small farm debate, arguments are complex and evidence in the form of quantitative data frustratingly scarce. The 62,827 farms of less than 50 hectares in England in 2016 accounted for over half (59 per cent) of all farm holdings in England (Defra, 2018). Yet despite being numerically important, at the risk of stating the obvious, they are small and consequently account for only 12 per cent of all farmed land. This suggests that the contribution to aggregate agricultural output from this group of farms will be limited. However, the relationship between farm size and output is complicated and influenced by efficiency and intensity of production. Focusing specifically on cereal farms as part of an Agricultural Change and Environment Observatory Research Report, Defra (2011) modelled the relationship between farm size and expected agricultural output based on a standard £150,000 of inputs per year for each of three farm sizes. The results presented in Table 6.4 appear to demonstrate a clear and positive association between farm area and the value of agricultural output. In other words, more value for the same level of input is achieved on larger farms. However, the relationships underlying these results are complex. For example, as Defra point out, although these modelled results are for cereal farms, they include all of the enterprises on the farms and these may be influencing the results.

Defra go on to warn about the complex relationship between farm size and efficiency. Indeed, there is a longstanding debate concerning the association between farm size and efficiency, complicated by different approaches to

Table 6.4 The relationship between farm size* and predicted output (cereal farms only)

	Land area (ha)		
	150	*200*	*250*
Estimated Farm business output (£'000s)	203	219	231
Estimated agricultural output (£'000s)	132	140	146

Note: *Farm size was measure in ESUs – European Size Units – where one ESU is defined as 1200 European Currency Units of Standard Gross Margin. ESUs therefore provide a measure of the economic size of the business. Farms were categorised as: very small =<8 ESUs; small 8-<40 ESUs; medium 40-<100 ESUs; large 100-<200 ESUs; Very large 200 ESUs and over. Farms of less than 8 ESUs were considered (in the UK) to be below the threshold of full-time activity.

Source: Defra, 2011, p.15

measuring efficiency as well as different ways of measuring farm size. Using data from the Farm Management Survey (the precursor to the FBS), the Zuckerman Committee (1961) concluded that, with the exception of arable farms, smaller holdings used significantly more inputs per unit area than larger farms of the same type (with the implication that they were less efficient). This finding may be a reflection that farmers consider detailed input calculations less worthwhile because of the modest quantities involved on a small farm (Gasson, 1988). However, the seminal study by Britton and Hill (1975) showed that, measuring farm size in standard man days, there were only small differences in the value of inputs per acre between farm size groups, although smaller farms were found to be more labour intensive (as we have seen above). More recently, Defra's (2011, p.25) results broadly suggest that 'there is an underlying tendency for larger farms (in financial terms) to be slightly more economically efficient than smaller ones', but this is only apparent when 'compounding factors' such as unpaid labour (charged at the full economic rate) are included. Furthermore, there is a *very large* level of variation within this relationship and the top performing small farms are more efficient than *many* larger farms.

One reason for this excursion into the complexities and intricacies of the farm size-intensity-efficiency relationship is to demonstrate that there is no simple, unequivocal answer. Proponents of both small and large-scale farming should bear this in mind when making claims regarding the efficiency of different-sized farms. Furthermore, the contribution of small farms to agricultural output is not just about efficiencies and their proportionate contribution to aggregate output. For instance, as we explore below, small farms are seen to make a particular contribution to rural society and communities.

The social, cultural and community contribution

The role of small family farms in rural society provides some of the most powerful imagery and assumptions in support of the contribution of small farms. Discussion of the social and community contribution is often powerfully normative but it is an area that remains underresearched. The exact nature of the 'social' dimension is also not always clear but can range from the role of farms in creating employment, to helping to sustain rural services and community institutions, through to the personal benefits of working on a small farm and the contribution of the operators of such farms to the 'national character'. So, for writers such as Wendell Berry (1987, p.347), the family farm is 'part of the definition of one's own humanity', supporting a superior quality of life and higher moral and spiritual values than industrial society.

As we saw in Chapter 1, in an address to the 2018 Oxford Farming Conference, Secretary of State for Environment, Food and Rural Affairs, Michael Gove, referred to the importance of smaller farms to rural communities and rural culture. Not only is it very unusual for a UK government minister to comment on the value of a particular size of farm, but to combine this with reference to culture and communities speaks to the notion that family farming still has an important place within wider society. This is reflected in the powerful appeal of books such as The Shepherd's Life (Rebanks, 2015) and the continued attraction of small family farming as an alternative to large-scale corporate farming, as reflected, for example, in the emergence of farmers' markets (see Chapter 4). In general, small family farms are seen to contribute to the spiritual and moral fabric of society: 'Farming is a way of life. Farmers contribute more than food to the welfare of the nation, in terms of ability, character, morality, work habits, experience of the natural world' (Weiss and Wilson, 1991, quoted in Wilson, 1996, p.237). More recently, Pretty (2013, p.108) has argued that 'social connectedness, trust and participation in community life was greater where farm scale was smaller'. Newby et al. also highlight the link often made between the supposed characteristics of small family farmers and the nation's health:

> The yeoman virtues of sturdy independence and solitary self-help have long been prized and celebrated as a source of strength in the English national character. … It is worth noting that this perspective continues to infect much of the thinking and writing on what has come to be known as the 'small farm problem'.
>
> (Newby et al., 1981, p.38)

Moving beyond assertion, what evidence there is regarding the contribution of small farms often comes from broader studies into social change in farming and rural communities. For example, in their research into East Anglian farmers in 1972, Newby et al. (1978) recognised that what had previously been farmers' 'natural place' in the local community was increasingly uncertain due to an influx of 'newcomers/outsiders/aliens' to the village, who were not dependent for their employment on local farmers and therefore undermined the ability of farmers and landowners to dominate whole communities in the way they had been used to. The 'cultural competences' (Cloke et al., 1998) of these newcomers, who have been able to take the place of farmers and assert their own values, has meant country living has become more similar to suburban territory, bringing with it new types of cultural conflict and implications for farmers. Parry et al. (2005, p.65) contend that 'the traditional mainstays of rural and farming life – the pub, the church and markets (are) in widespread decline, partly because of competing time pressures on farmers, and partly because of the changing nature of the rural population'. The implications of social change in rural communities that Newby et al. (1978) began to identify were exacerbated by a lack of appreciation and understanding among the wider community.

Such trends have continued. Reed et al. (2002, p.38) describe a 'collapse of solidarity' in rural communities, where farmers played an 'important but limited civic role in the broader community'. Similarly, Burton et al. (2005) and Appleby (2004) identify a decline of 'social capital' in UK farming due to the erosion of traditional community ties and working arrangements. In their investigation into the wider social impacts of agricultural restructuring in 2005, Lobley et al. also identify farmers' withdrawal from rural society and decline in 'social connectivity'. They argue that:

> Despite being socially embedded in their communities (that is living very near their place of birth and most of their close family and friends) the results of the household survey suggest that farmers are less socially active than non-farmers. The reasons for this vary but are associated with a desire to avoid exposure to criticism (of farming/being a farmer), the lack of time associated with excessive working hours and, more straightforwardly, the declining number of main occupation farmers in rural areas.
> (Lobley et al., 2005, p.vi)

As we discuss further in Chapter 7, farmers' changing place in the community has had profound implications. Withdrawal from the community can lead to a downward spiral of depression, stress, illness and difficulties within interpersonal relationships; which have obvious implications for the farm business. Conversely, according to Meert et al. (2005), social interaction and integration into social networks are significant factors in a farmer's decision to establish new diversified enterprise; that is, farmers' lack of social interaction, as reported by a number of commentators across a range of contexts (Appleby, 2004; Burton et al., 2005; Lobley et al., 2002; 2005), has implications for farmers' propensity to establish new enterprises as well as direct implications for their own well-being. Thus, the 'social' is inseparable from the economic.

The picture regarding farmers' role in the community is not entirely gloomy, however. Participants at a stakeholder workshop conducted for Winter and Lobley's (2016) research argued that small farmers continue to play an important community role, either formally on Parish Councils or informally, as the people in the village turn to them to remove fallen trees that are blocking the road, or to stack firewood for bonfire night, for instance. Findings from the 2015 SIP Baseline Survey also suggest that most farmers are still involved with their local community to some extent, with many making a notable contribution. Of small (<50ha) farmers in the sample, 94.8

per cent said they had contact with non-farming members of the local community at least once a week. While not statistically significant, fewer numbers of large and very large farmers reported this level of contact (75.8 per cent and 82.3 per cent respectively), suggesting that the operators of small farms may have stronger links to their local area – though the proportion of larger farms also reporting regular contact remains encouraging. Despite not being directly asked about community activities they were involved with, 29 per cent of all farmers mentioned participating in at least one community-focused group or activity. These activities were numerous and varied but included membership of the parish council, local sport clubs, church-related activities, singing groups, Rotary clubs and general village activities such as helping out at village fetes (e.g. providing straw or hay bales for props). Some farmers even talked at length about how they still perceive farming to be the 'backbone' of the local community. For instance;

> We think that farming is essential, we think it provides the backbone and a framework. We've got a little tiny village and people come and people go, but when you want to get anything done, when there's money to be raised for the church, the community – we're involved with all of it – it's always, new people come in and have got new ideas, and it's great because it provides enthusiasm, but when it becomes boring and it's old hat, it's the farming families that go on doing it and do all the boring bits.
>
> (LFA livestock farmer, Taw catchment)

Findings from the 2016 SW Farm Survey also point to a potential relationship between farm size and community embeddedness (although these findings were not statistically significant). The survey found that small famers of under 50 hectares were more likely to 'strongly agree' with the statement that 'farming is essential to the local community' than larger farms (see Table 6.5). Again though, levels of agreement with this statement by larger

Table 6.5 Agreement with the statement 'Farming is essential to the local community', by farm size (%)

	Farm size (ha)						*All farms*
	<50	*50<100*	*100<150*	*150<200*	*200<250*	*250+*	
Strongly disagree	6.7	6.1	10.8	5.1	8.0	8.7	7.4
Disagree	10.5	7.9	10.8	9.1	12.0	12.8	10.1
Neither disagree nor agree	14.7	23.1	18.1	22.2	24.0	21.5	19.8
Agree	19.2	24.3	21.6	28.3	12.0	20.1	21.5
Strongly agree	48.9	38.6	38.7	35.4	44.0	36.9	41.2

Note: χ^2*=30.12, p<.0068 (not significant).*

Source: 2016 SW Farm Survey

farmers were also generally high; a reminder that larger size is not necessarily a barrier to engagement with the local community.

Another piece of evidence that suggests that small farmers might have closer social ties to their local community than larger operators is provided by Milbourne et al. (2001). In an analysis of verbal or written complaints to farmers from members of the public or representatives of public agencies, the authors observed a very strong relationship between levels of complaints received by farmers and farm size (see Table 6.6). It is striking indeed that 72 per cent of farmers with over 500 hectares experienced complaints, whereas 78 per cent of small farmers (less than 50 ha) had received no complaints at all. Thus, in a number of categories where there are few complaints overall, larger farmers are significantly affected. Important examples include: other smells (16 per cent); health risks from pesticides (11 per cent); destruction of wildlife or landscape features (14 per cent); and plans to sell land for development (11 per cent).

In addition to general contributions to the social life of their local community, there are some examples of where small farms have quite specific important social and community benefits. For instance, care farming initiatives,[2] which tend to be based on small farms,[3] have been shown to have significant benefits for the physical and mental health of participants (Bragg et al., 2014; Hine et al., 2008; Leck et al., 2014). Other small farms actively encourage members of their local community to visit the farm and provide educational visits for school children and/or organise activities such as community gardening and family-oriented open days. Examples of such farms include Grown Green @ Hartley Farm in Wiltshire (https://growngreen.wordpress.com) and West Town Farm in Devon (www.westtownfarm.co.uk).

A final example of where small farms explicitly provide a link between people and the land and producers is Community Supported Agriculture (CSA), where the risks and benefits of farming are shared between farmers and CSA members. Although more popular in the US where it has been established since the 1980s (and where most CSA research has been focused), the number of CSA initiatives in the UK is growing, with over 100 currently listed on the CSA network's website (https://communitysupportedagriculture.org.uk/). CSA enables those who consume the food (the members) to support the farmer by committing to buy a 'share' in the harvest at the beginning of the season, and maybe contributing to work on the farm when the labour demand is high. The aim is to establish a mutually beneficial relationship between farmers and consumers, and to build up a community around the farm. Hence, many CSAs also provide their members with social benefits such as farm picnics, bonfires and an annual dance. Research has shown that, while not entirely unproblematic, CSAs can be successful at fostering connections between farmers and members, strengthening appreciation for food and farming, raising environmental consciousness and improving health and well-being (Brown and Miller, 2008; Hayden and Buck, 2012; Saltmarsh et al., 2011).

Table 6.6 Incidence of verbal or written complaints about activities from members of the public or representatives of public agencies, by farm size in GB (% of farmers)

	Farm size (ha) (crops and grass)					
	<49.9	*50–99.9*	*100–199.9*	*200–299.9*	*300–499.9*	*>500*
No complaints received	78	72	61	53	60	28
Public access issues, such as footpaths	6	10	17	23	22	28
Mud or slurry on roads	5	8	16	24	13	44
Hedge trimming	4	4	5	6	7	5
Use of slow moving heavy machinery on roads	2	3	4	5	9	9
Smell from livestock units	2	2	2	2	-	-
Other smells	2	2	2	3	2	16
Health risks to public from livestock units	-	-	1	1	-	-
Health risks to public from pesticides	-	1	2	1	-	11
Flies or vermin believed to emanate from farm	3	-	1	3	-	-
Proposed or recently constructed agricultural buildings	-	1	1	2	6	5
Pollution or contamination of water courses	1	2	3	3	3	-
Agricultural 'litter' (such as silage bags etc.)	1	-	-	-	-	4
Destruction of wildlife and landscape features (incl. hedges)	-	1	2	2	6	14
Game shooting on your land	-	1	2	2	2	5
Hunting on your land	-	1	1	4	3	-
Residential barn conversions	-	-	1	-	-	-
New non-agricultural uses for farm buildings	1	-	1	-	-	-
Plans to sell land for development	1	2	1	3	2	11
Plans to use land for non-agricultural purposes	1	1	1	-	1	-

Source: Milbourne et al., 2001

The environmental contribution

Research into the environmental implications of farm size is limited, despite large bodies of literature pertaining to the contamination or degradation of the environment and surrounding ecosystems (e.g. damage to soil, leaching, run-off and eutrophication), and despite a longstanding anecdotal belief that large farms are the most damaging to the environment (Heffernan and Green, 1986). What evidence there is suggests, as always, a complex picture.

Although now increasingly dated, Potter and Lobley (1992, 1993) and subsequently Lobley (2000) provide the most comprehensive discussion of the environmental contributions of small farms in a British context. They recognise that what they term the 'strong conservation case' for small farms is often based on a belief that their operators are somehow inherently more sensitive to the environment and therefore provide the best way to organise and manage the land, but they also identify other arguments in support of the conservation value of small farms:

> If small farms can be shown to be more environmentally sensitive, is this because the people managing them are more conservation-minded or is it because they are conservationist by default – they lack the means to intensify production in the way a larger, more prosperous farmer might? Alternatively, are small farms conservationist by association because they tend to be of a type that is environmentally sensitive or rich in conservation assets anyway?
>
> (Potter and Lobley, 1993, p.271)

Thus there are two issues here. First, is there evidence that environmental performance or quality is related in any way to farm size? Second, if that is in some way the case, why is this so – is it by inclination (i.e. the attitudes and behaviour of the operators of small farms), default (i.e. the inability or unwillingness of small farm operators to engage in actions that have a negative impact on the farmed environment) or association (i.e. are small farms associated with types of agriculture that are naturally rich in environmental assets)?

Contributors to Winter and Lobley's (2016, p.49) call for evidence certainly shared the view that small farmers were more inclined to farm in a manner that is environmentally friendly:

> Small farms are often more willing to look at their environmental impact and more willing to assign areas of the farm to wildlife. Small farms tend to be more sensitive to the land they occupy and farm in a way which more fits the landscape. Small farms will often be less intensive in their use of the land and be more sustainable for the environment.
>
> (Joel Woolf, Partner, Head of Agriculture, Foot Anstey)

> Small farms also act as magnets for wildlife that exists at low population densities on their larger neighbours. Small farms are brilliant for monitoring wildlife that exists invisibly on larger areas. On our little farm we have different habitats from our immediate neighbours and have locally unique species. However, we know that some species have a meta-population dynamic that means that they must exist on our neighbours land even though they are invisible there.
>
> (Huw Jones, Glyn-Coch Farm)

The available empirical evidence suggests a more complex situation. Evidence from a survey of 504 British farmers in 1993, using the concept of conservation capital[4] (1997; 2000), demonstrates that smaller farms (<50 ha) are more likely than larger farms (>200 ha) to have zero stock of conservation capital (39 per cent versus 23 per cent); see Table 6.7. However, by separating the small from the very small, the latter, while still containing a significant proportion of zero stock parcels (33 per cent), emerged as having the highest concentration of high conservation stock parcels across the sample (50 per cent).

While this suggests very small farms are of high conservation value, they only account for 10 per cent of the stock conservation capital in the survey (because they represent a much smaller proportion of the total land area), while the largest farms (200 hectares or more) account for 59 per cent. However, this distribution does vary by landscape type[5] (see Table 6.8). For example, in arable landscapes 5 per cent of conservationist capital is located on farms of less than 50 hectares, compared to 24 per cent in pastural[6] landscapes, suggesting that location and landscape factors are important in determining conservation value, as well as farm size. These results point to a balanced conclusion about the value of small farms to the environment, suggesting that

> there is some support for the assumption of conservation interest by association [and] in some locations (notably pastural landscapes) they [small farms] are responsible for a significant proportion of conservation

Table 6.7 Degree of conservation capital by farm size (%)*

Conservation capital	*Very small <20ha*	*Small 20–50ha*	*Medium 50–200ha*	*Large >200ha*	*All farms*
Zero	32.8	44.3	32.5	22.7	31.8
Low	17.2	2.6	40.1	37.9	34.1
High	50.0	29.1	27.4	39.4	34.1

Note: *The association between farm size and conservation capital was significant at 0.00026 level using Chi-square.

Source: Lobley, 2000a

Table 6.8 The distribution of conservation capital by landscape type and farm size (% of area of conservation capital)

Landscape type	*Very small <20ha*	*Small 20–50ha*	*Medium 50–200ha*	*Large >200ha*	*Total*
Arable	3.5	1.3	27.7	67.5	100
Pastural	9.1	14.7	3.3	39.9	100
Marginal and upland	12.1	1.7	14.9	71.3	100

Source: Lobley, 2000a

> capital, and their loss could expose land of conservation value to potentially environmentally damaging structural change
>
> (Lobley, 2000a, p.601)

Thus, while small farms only manage a small proportion of the land, they nonetheless play a fundamental role in the collective provision of rural environment services.

Assessing the environmental performance of farms is a complex task, however, with results depending on the particular indicators that are measured. Hence, there is some evidence that small farms may be no more – and in some respects less – 'environmentally friendly' than larger farms when analysed in terms of resource-use efficiency and overall environmental footprint. Using the Agri-Environmental Footprint Index (AFI)[7] in combination with Farm Business Survey data for arable, lowland livestock and upland livestock farms, Westbury et al. (2011) found that farm size (measured in terms of land area) had no significant effect on AFI values calculated for arable and upland livestock farms. However, on lowland livestock holdings farm size had a significant effect on AFI, with a significant increase in environmental performance with farm size. Westbury et al. (2011) attribute this positive relationship to larger holdings using significantly less energy per hectare than smaller holdings, as well as having greater land use diversity and using less water use per hectare compared to their smaller counterparts (Table 6.9).

Other work also points to the dangers of drawing straightforward conclusions about the relationship between farm size and environmental performance. With reference to farmers' behaviour towards water quality management in Nitrate Vulnerable Zones (NVZs) in Scotland, Barnes et al. (2011) utilised cluster analysis to identify three distinct groups according to their agreement with a number of key statements: 'resistors', 'multifunctionalists' and 'apathists'. Smaller farmers were most likely to be apathists and appeared to be generally disengaged with the regulations; but it was the largest farmers (in both income and acreage terms) who had the most negative views ('resistors'), disputing the link between water quality and the health status of the farm, and tending to avoid responsibility for water quality. Despite this, resistors

Table 6.9 Indicator values for lowland livestock holdings according to farm size

	Small *<80ha* *(n=60)*	*Medium* *80–120ha* *(n=27)*	*Large* *>120ha* *(n=51)*
Fertiliser units (tonnes) per ha	0.59	0.34	0.50
Grazing livestock units per ha forage	1.89	1.42	1.59
Energy consumption – units per ha	90.48	97.49	60.31
Water usage m³ per ha	33.28	28.36	18.47
Rough grassland % of utilisable agricultural area	2.96	0.04	3.05
Temporary grassland – % of total grassland	27.96	12.41	34.25
Woodland cover % of total farm area	0.19	0.13	2.49
Land use diversity	0.27	0.41	0.60

Source: Reprinted from the *Journal of Environmental Management*, Vol. 92, Westbury, D. B., Park, J. R., Mauchline, A. L. Crane, R. T. and Mortimer, S. R. Assessing the environmental performance of English arable and livestock holdings using data from the Farm Accountancy Data Network (FADN), pp.902–909, Copyright (2011) with permission from Elsevier

had the highest rate of adoption of water management practices – something which Barnes et al. (2011) attribute to the group's emphasis on resource management – and were the most likely to take advice from consultants and advisors from outside the business. Notably, the smaller farm-sized 'apathists' demonstrated least change in their nitrogen management behaviour.

In a similar vein, Ingram (2008) observed how small farms failed to see manure as an asset, in contrast to bigger farms, which were 'more disciplined about accounting for manure, measuring its value as part of their nutritional programme and using more sophisticated spreading machinery' (Ingram, 2008, p.221). Nutrient budgeting was seen as something that only larger arable farmers took an interest in, with smaller farmers often being constrained by the size of farm, poor soils and their own lack of experience. On the other hand, there was some suggestion by farmers and farm advisors that the use of large machinery – typically on larger holdings – is threatening farmers' knowledge of the soil by removing 'their physical and sensual contact with the soil, obscuring any visible signs of problems with the subsoil, which may have been detected earlier by someone on foot' (Ingram, 2008, p.223). This issue was felt to only be relevant for larger farmers who hire labour and utilise large machinery.

Economics of contemporary small farms

As we have already seen, powerful economic forces continue to force the pace of farm growth. The infamous agricultural treadmill means that ever larger volumes of output are needed just to stand still in net income terms. Clearly, however, not all farms are growing in size. As discussed in Chapter 3 and earlier in this chapter, some farmers pursue alternative strategies to generate income, through diversification and adding value. Others are supported by income generated off the farm and others adopt the tried and tested strategy of tightening the belt, although there is a limit to how far the belt can be tightened.

Drawing primarily on analyses of Farm Business Survey (FBS) data by Wilson (2016), this section examines the underlying economic circumstances of small farming and seeks to establish the factors that are associated with successful small farm businesses. This provides a more detailed analysis of some of the issues covered in Chapter 3, focusing on small farms (defined according to SLR in this instance). The FBS analysis is of data from 2418[8] observations from the 2014/15 England and Wales FBS (for full details see Wilson, 2016). Note that the data were weighted drawing upon the standard FBS weights in order to produce national (England and Wales) estimates. Clearly, being based on a single year, the financial figures presented here are in part influenced by factors such as agricultural commodity prices, input costs and exchange rates at the time. Nevertheless, although figures may change from year to year, the relationship between different categories of farm remains instructive.

Farm Business Income (FBI) represents the financial return to the farm business and is Defra's preferred measure of farm business performance. Figure 6.3 shows FBI results for England and Wales by farm size and EU region. It is no surprise that the largest FBI is achieved by large farm

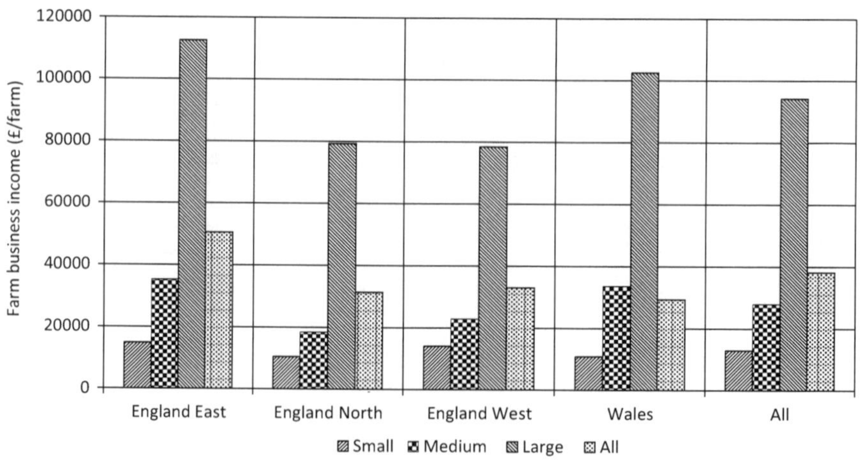

Figure 6.3 Farm Business Income (£/farm) by farm size and EU region

Note: * based upon Standard Labour Requirements (SLR). Small = <2 SLR; Medium = 2-<3 SLR; Large = 3 SLR or greater.

Source: Wilson, 2016

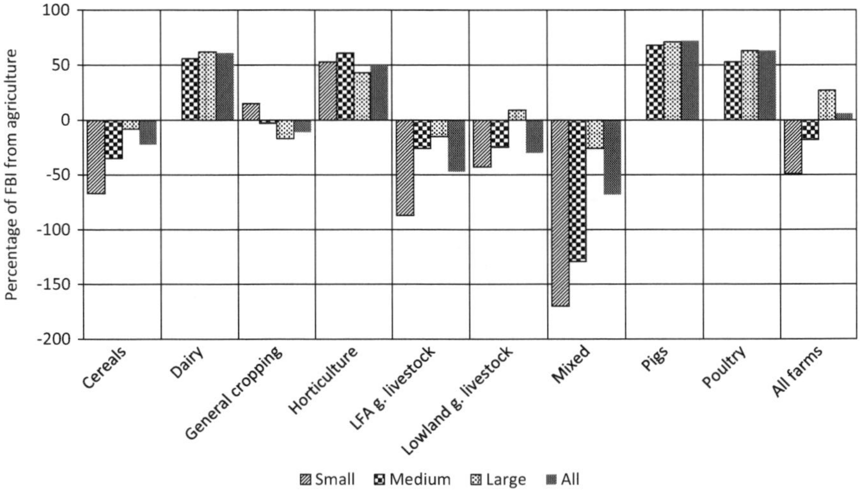

Figure 6.4 Percentage of FBI derived from agriculture, by farm type and size

Source: Wilson, 2016

businesses in the east of England, with the smallest FBI being recorded by the small farm size category in each region (Wilson, 2016).

The FBS collects data on the sources that contribute to FBI, which is helpful in identifying the relative dependency on different income sources and areas of possible vulnerability. Figure 6.4 indicates the proportion of FBI that is derived from agriculture. It can be noted that, as a group, both small and medium-sized farms make a loss on their agricultural account.

Figure 6.5 demonstrates the significance of agri-environmental payments. It can be seen that the relative contribution of such payments declines with increasing farm size and that, for small mixed and less favoured area (LFA) grazing livestock farms, agri-environmental payments make a substantial contribution to overall FBI. Turning to diversification, Figure 6.6 indicates that, like agri-environment income, the contribution of diversification income to overall FBI generally declines with increasing farm size. The importance of diversification as an income source for small mixed, lowland livestock and cereals farms is particularly clear. Finally, Figure 6.7 demonstrates the importance of the Single Farm Payment (SFP) (the precursor to the Basic Payment Scheme) to all farms, but particularly small and medium-size farms. It is also notable that the SFP contributes in excess of 100 per cent of FBI on small mixed farms.

The above analysis clearly demonstrates that small farms are dependent on income sources other than those derived from the sale of crops and livestock. The significance of the SFP to the FBI of both small and medium farms shows how vulnerable they could be to a significant reduction, or even loss, of the payment. Data from the 2016 SW Farm Survey confirms the importance

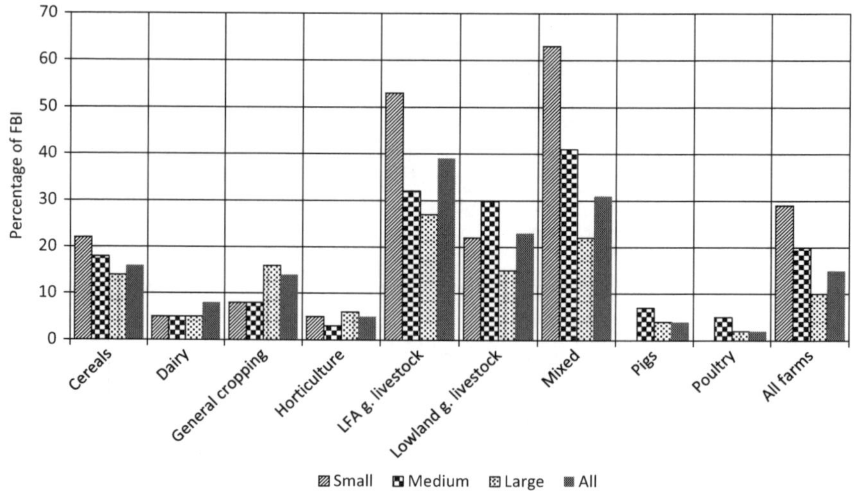

Figure 6.5 Percentage of FBI derived from agri-environment payments, by farm type and size

Source: Wilson, 2016

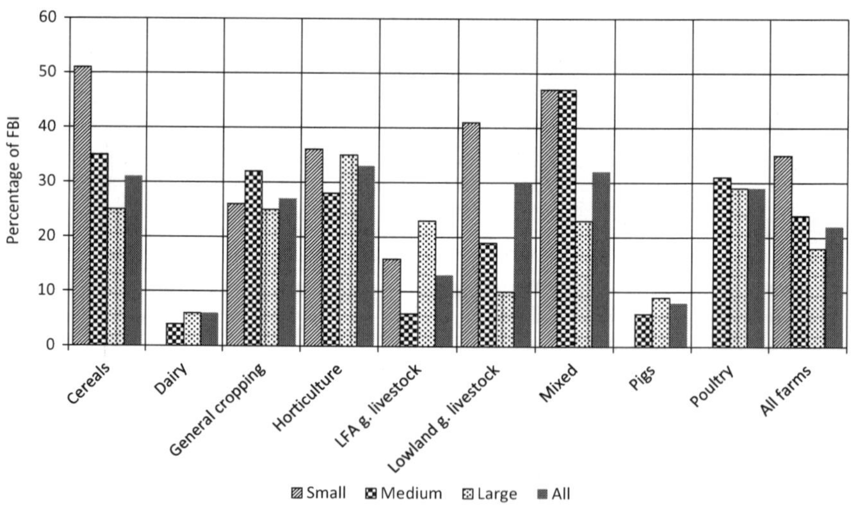

Figure 6.6 Percentage of FBI derived from diversification, by farm type and size

Source: Wilson, 2016

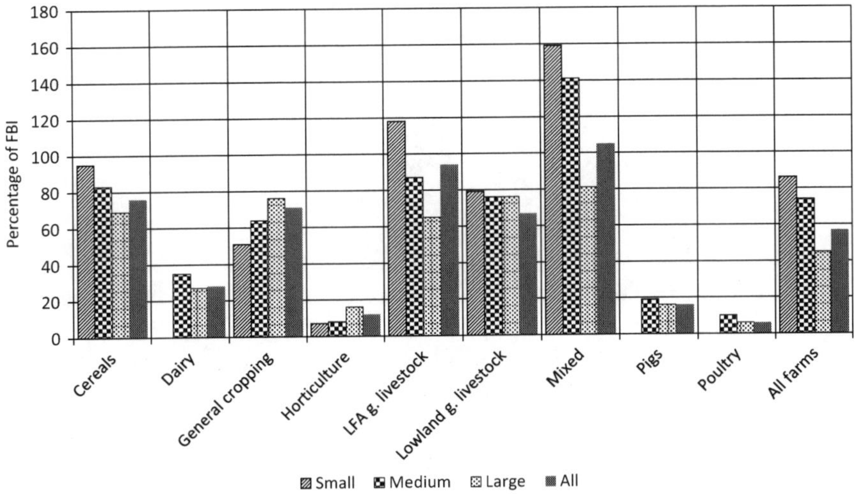

Figure 6.7 Percentage of FBI derived from SFP, by farm type and size

Source: Wilson, 2016

Table 6.10 Mean proportion of income from different sources, by farm size (%)

	Farm size (ha)						*All farms*
	<50	*50<100*	*100<150*	*150<200*	*200<250*	*250+*	
Agriculture on this farm	43.7	61.9	71.3	75.5	76.4	73.0	62.2
Non-agricultural enterprises on this farm	15.5	11.5	11.8	11.1	9.7	9.5	12.2
Income from off-farm work	13.4	8.9	5.4	7.4	6.1	6.3	8.8
Pensions, savings, investments	20.0	13.7	7.3	4.6	4.0	8.5	12.2

Notes: Chi-squared values are as follows: Agriculture χ^2*=178.65, p<.001;* Non-agricultural enterprisesχ^2 *not significant;* Income from off-farm work χ^2*=39.80, p=.005 but 40 per cent cells have expected count less than 5;* Pensions etc. χ^2*=69.64, p<.001 but 30 per cent cells have expected count less than 5.*

Source: 2016 SW Farm Survey

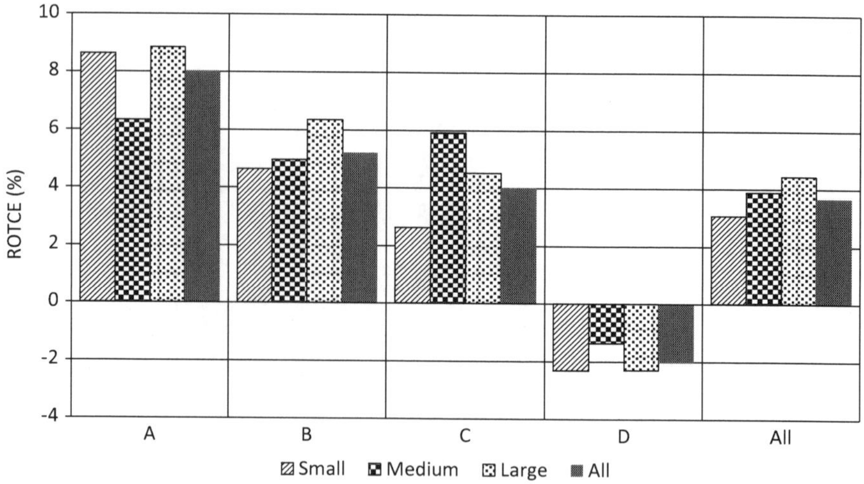

Figure 6.8 Return (FBI) on ROTCE by FBI performance group and farm size

Source: Wilson, 2016

of non-agricultural income for small farms (measured here according to land area rather than SLR), with the smallest farms (< 50 ha) deriving on average just 44 per cent of household income from agriculture, compared to 73 per cent for the largest farms (250+ ha) and 62 per cent for the sample as a whole (see Table 6.10).

Wilson (2016) also reports on a number of other measures of financial performance and their association with farm size. Focusing just on the agricultural element, Wilson looks at the value of output for every £1 of cost (including an imputed value for unpaid family labour). On average across the FBS sample, for every £1 of cost an agricultural output of £0.77 was achieved; ranging from £0.64 for small to £0.96 for large farm businesses.

Figure 6.8 illustrates the financial returns to the business measured by FBI as a proportion of tenant's type capital[9] employed (ROTCE) in the business. It is worth noting that, not only do the best small farms gain a better return than medium farms, this is broadly in line with the returns on large farms.

Another measure of financial performance and potential business vulnerability is gearing (total liabilities as a percentage of net worth). As Figure 6.9 indicates, this is lower on small farms across all FBI performance quartiles. In other words, small farms have a lower level of debt relative to their overall business worth than larger businesses, indicating the potential for a greater degree of financial stability (Wilson, 2016; also see Andersons, 2016).

The combination of FBS data and analysis of the SW Farm Survey provides valuable insights into the contemporary economics of small farms and points to the heterogeneity of the small farm sector. Wilson (2016, p.46) states that

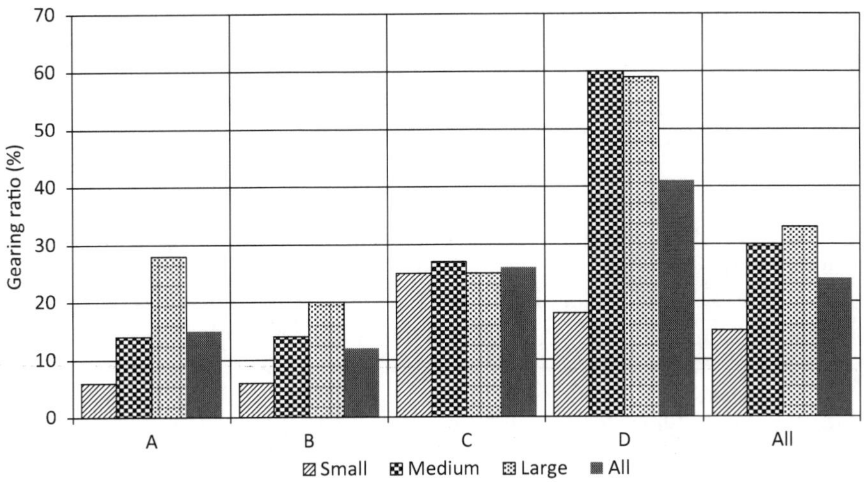

Figure 6.9 Gearing ratio, by FBI performance group and farm size

Source: Wilson, 2016

> irrespective of farm size, profitable farm businesses are underpinned by a profitable agriculture cost centre. Moreover, these businesses typically also achieve greater levels of income from agri-environment, diversification and SFP sources. It is, however, important to note that successful farm businesses rely least (in percentage terms) on the SFP as a source of income.

In other words, profitable and successful farm businesses are good at everything they do.

Small farms, in general, are associated with more modest FBI – why would we expect otherwise – but as we have seen, this is often supplemented by other income from off the farm (both earned and transfer payments) and small farms are associated with a favourable return on capital. This, combined with an equally favourable debt to asset ratio, suggests that there is a platform from which to maintain a sustainable business or expand. Of course, not all operators of small farms will wish to expand. The heterogeneity apparent throughout the analysis presented here reflects a range of different types of small farm ranging from what may be considered 'main living' small farms through to 'lifestyle' and 'retirement' holdings.[10] Some, with little or no debt and owning their own land, may be content to 'absorb' adverse changes in the economics of agriculture by adopting an ever more frugal lifestyle or supplementing with income from elsewhere (see Lobley et al., 2005, for a discussion of 'absorbers').

Other small farmers need to make changes to their production and/or business practices in order to improve the current generally unfavourable output-input ratio. One way of doing this is to take control of the supply chain by selling direct to the consumer. Although this is a fairly common practice on organic farms, there appears to be scope for wider take-up of this strategy by small farms generally: only 8 per cent of farms under 50 hectares (and 7 per cent of all farms) in the South West were engaged in the processing/retailing of farm produce in 2016 according to our SW Farm Survey data. In addition to various strategies for capturing a greater share of the value of a farm's output, Wilson's (2016) FBS analysis also identified the potential gains in small farm competitiveness that could be achieved through building on what are currently low levels of labour and/or machinery sharing.

Finally, as Wilson (2016, p.36) states, 'characteristics of farm succession arguably represents one of the largest risks for the future viability of family farms', although he goes on to note that the most profitable farms are the most likely to have a successor, regardless of farm size. The relationship between succession and farm size, and indeed business performance, is, however, complex. To an extent the lack of a successor on some small farms should not be seen as a problem. Those that are effectively retirement holdings of one variety or another should not be expected to attract a familial successor. Whether they might ultimately offer an opportunity for a new entrant is another matter. More problematic are the small farms where there is a desire for intergenerational succession within the family, but where the business cannot support an additional salary during the transitional stage (which may take years). If they are unable to grow, or improve their output-input ratio or alter their business model so that they capture a greater proportion of the end value of their produce, or develop new income streams, these are the farms that are inter-generationally vulnerable.

Conclusions: the distinctive contribution of small farms

Advocates of small farms make a strong case for the positive contribution that small farming makes to rural life and the countryside. As this chapter has demonstrated, evidence, in the form of large-scale statistical surveys for instance, is uneven but there are many examples of the role played by small farms. However, the evidence that is available typically suggests greater complexity than simply being able to point to a clear-cut relationship between farm size and environmental value, just to take one example. Rather, it appears to be the complex interplay between size, farm type, attitudes and dispositions to behave in certain ways that contributes to the role played by small farms.

In addition to extolling the virtues of small farming, proponents point to the consequences of a decline in the number of small farms. The loss of small farms, it is argued, is associated with fewer people on the land and fewer to play formal or informal roles in communities. As we have seen, there is evidence that famers have already withdrawn from various community roles. Further declines in the number of small farms would probably mean fewer local suppliers of food and other services. The environmental implications would depend very much on what replaces small farms and it would be just as dangerous to assume that all large farms are environmentally damaging as it would to assume that all small farms are environmentally beneficial. Ultimately, rather than privileging one set of farm structures over another, it is a question of maintaining a diversity of farm size structures. This of course depends on the ability of small farms to be economically viable.

The infamous agricultural treadmill means that ever larger volumes of outputs are needed for farm businesses just to stand still in net income terms. We have also seen just how dependant many small farms are on the current Basic Farm Payment and how vulnerable they could be to a significant reduction, or even loss, of this payment. That said, heterogeneity apparent throughout the analysis presented here means that some will continue to absorb change while others will need to adopt various strategies to capture a greater share of the value of the farm's output and/or achieve scale effects by collaborating with other farmers, for example, via labour and/or machinery sharing. We will return to some of these issues in the concluding chapter.

Notes

1 The study areas were The Peak District: Bakewell area; The High Weald: Heathfield area; East Midlands: Newark area; Cumbria: Orton Fells area; Mid Devon: Witheridge area; North Norfolk: Fakenham area.

2 Care farming is defined by the National Care Farming Initiative (now Care Farming UK) as 'the therapeutic use of farming practices' (RDPE Network and National Care Farming initiative, 2010, p.4), and is based on involving socially, physically and/or mentally disadvantaged individuals in normal agricultural practices on commercial farms.

3 The average size of care farms in the UK in 2008 was 49.9 hectares (Hine et al., 2008).

4 Conservation capital is a measure of the area of deciduous woodland, semi-natural vegetation and extensive grass. Data from Countryside Survey 1990 was used to measure the extent of conservation capital different 'occupier parcels' (defined as the area of a farm included in the CS1990 field survey).

5 The landscape type classification was derived through the combination of individual land classes into distinct groups. Although each landscape has a distinct spatial distribution, assignment to a particular landscape does not imply anything about specific location (e.g. arable landscapes can be found in the East Midlands and South of England). A detailed description of each landscape type is provided by Barr et al. (1993) (also see, Bunce and Howard, 1992). Some of the characteristics of the landscapes referred to in the following analysis are as follows:

Arable landscapes – dominated by arable crops and intensive grass, found largely in Southern England, East Anglia and East Midlands, but also in parts of the East Coast of Scotland.

Pastural landscapes – characterised by large areas of grassland, small fields, hedgerows and small woods. This landscape is typical of South West England, West Wales, parts of the Welsh/English border, and the North West.

Marginal and upland landscapes – found in northern Britain, Wales and Scotland. Dominated by a mix of low-intensity livestock farming and forestry. This landscape contains extensive tracts of semi-natural vegetation.

6 The spelling of pastural is taken from CS90 and should not be mistaken for pastoral. The name is derived from the dominant characteristics of the landscape, i.e. pasture.
7 A farm-level measure of environmental performance which aggregates a range of measurements of agri-environmental indicators; a high score depicts good environmental performance.
8 Smaller sample sizes were used for certain specific data analysis, specifically in relation to non-farm business income sources (1,495) and the presence or absence of a nominated successor (2,418).
9 Closing valuations for: machinery, livestock, glasshouses, permanent crops, crops, forage, cultivations, stores, liquid assets, and Single Payment Scheme entitlements.
10 It is useful to note that within the US Department of Agriculture (USDA)'s definition of 'small family farms' (gross sales less than $250,000) the following distinct types are recognised: retirement farms; residential/lifestyle farms; farming-occupation farms; low-sales farm; and high-sales farms.

References

Akram-Lodhi, A. H. & Kay, C. (2009). *Peasants and Globalization: Political economy, rural transformation and the agrarian question.* Abingdon: Routledge.

Andersons (2016). *The Cash Flow Crisis in Farming and its Implications for the Wider Rural Economy*. London: The Prince's Countryside Fund.

Appleby, M. (2004). *Norfolk Arable Land Management Initiative (NALMI). Final Project report*. London: Countryside Agency.

Barnes, A. P., Willock, J., Toma, L. & Hall, C. (2011). Utilising a farmer typology to understand farmer behaviour towards water quality management: Nitrate vulnerable zones in Scotland. *Journal of Environmental Planning and Management*, 54(4), 477–494.

Barr, C., Bunce, R. G. H., Clarke, R., Fuller, R., Furse, M., Gillespie, M., Groom, G., Hallman, C., Hornung, M., Howard, D. & Ness, M. (1993). *Countryside Survey 1990, Main report.* CS 1990 Volume 2. London: Department of the Environment.

Berry, W. (1987). A defense of the family farm. *In:* Comstock, G. (ed.) *Is there a Moral Obligation to Save the Family Farm?* Ames: Iowa State Press, 347–360.

Berry, W. (2002). *The Art of the Commonplace: The agrarian essays of Wendell Berry.* Berkeley, CA: Counterpoint.

Bonanno, A. (1987). *Small farms: Persistence with legitimation.* Boulder, CO: Westview.

Bragg, R., Egginton-Metters, I., Elsey, H. & Wood, C. (2014). *Care Farming: Defining the 'offer' in England.* Natural England Commissioned Reports, No. 155. London: Natural England

Britton, D. & Hill, B. (1975). *Size and Efficiency in Farming.* Farnborough: Saxon House.

Brookfield, H. & Parsons, H. (2007). *Family Farms: Survival and prospect: A world-wide analysis.* Abingdon: Taylor & Francis.

Brown, C. & Miller, S. (2008). The impacts of local markets: A review of research on farmers markets and Community Supported Agriculture (CSA). *American Journal of Agricultural Economics*, 90(5), 1298–1302.

Brown, J. (2000). Agricultural policy and the National Farmers' Union, 1908–1939. *In:* Wordie, J. R. (ed.) *Agriculture and Politics in England, 1815–1939.* London: Palgrave Macmillan, 178–198.

Bunce, R. & Howard, D. (1992). *Aggregation of ITE Land Classes for Great Britain into Broad Groups*. Merlewood: Institute of Terrestrial Ecology.

Burton, R., Mansfield, L., Schwartz, G., Brown, K. & Convery, L. (2005). *Social Capital in Hill Farming. Report for the Upland Centre*. Aberdeen: Macaulay Institute.

Cleary, M. C. (1989). *Peasants, Politicians and Producers: The organisation of agriculture in France since 1918*. Cambridge: Cambridge University Press.

Cloke, P., Goodwin, M. & Milbourne, P. (1998). Inside looking out: outside looking in: Different experiences of cultural competence in rural lifestyles. *In:* Boyle, P. J. & Halfacree, K. (eds) *Migration Into Rural Areas: Theories and issues*. Chichester: Wiley.

Committee of Inquiry on Crofting (2008). *Committee of Inquiry on Crofting: Final report*. Edinburgh: Scottish Government.

Cox, G., Lowe, P. & Winter, M. (1991). The origins and early development of the National Farmers' Union. *The Agricultural History Review*, 39, 30–47.

Cragoe, M. & Readman, P. (2010). *The Land Question in Britain, 1750–1950*. London: Palgrave Macmillan.

Defra (2011). *Cereal Farms Economic Performance and Links with Environmental Performance*. Agricultural Change and Environment Observatory Research Report No. 25. London: Defra.

Defra (2018). *Structure of the agricultural industry in England and the UK at June*. [Online] Available at: www.gov.uk/government/statistical-data-sets/structure-of-the-agricultural-industry-in-england-and-the-uk-at-june [accessed 12/03/18].

Fennell, R. (1979). *The Common Agricultural Policy of the European Community: Its institutional and administrative organisation*. Oxford: Clarendon Press.

Flynn, A., Lowe, P. & Winter, M. (1996). The political power of farmers: An English perspective. *Rural History*, 7(1), 15–32.

Gasson, R. M. (1988). *The Economics of Part-time Farming*. Harlow: Longman Scientific & Technical.

Gosling, P. (2015). *On the Eve of Destruction: The case for Stormont intervention to save Northern Ireland's farming industry*. Report commissioned by Farmers For Action NI and the Northern Ireland Agricultural Producers Association. Online: Paul Gosling.

Hayden, J. & Buck, D. (2012). Doing community supported agriculture: Tactile space, affect and effects of membership. *Geoforum*, 43(2), 332–341.

Hayes-Conroy, A. (2007). *Reconnecting Lives to the Land: An agenda for critical dialogue*. New Jersey: Fairleigh Dickinson University Press.

Heffernan, W. & Green, G. (1986). Farm size and soil loss: Prospects for a sustainable agriculture. *Rural Sociology*, 51(1), 31.

Hine, R., Peacock, J. & Pretty, J. (2008). *Care Farming in the UK: Evidence and opportunities*. Colchester: University of Essex

Hoggart, K., Buller, H. & Black, R. (1995). *Rural Europe: Identity and change*. London: Edward Arnold.

Holden, P. (2015). If our small farms are allowed to wither, the whole nation will suffer. *The Guardian* [Online]. Available at: www.theguardian.com/commentisfree/2015/aug/09/lose-small-farms-britains-heritage-gone [accessed 7 December 2015].

Ingram, J. (2008). Are farmers in England equipped to meet the knowledge challenge of sustainable soil management? An analysis of farmer and advisor views. *Journal of Environmental Management*, 86(1), 214–228.

Kopsidis, M. (2012). *Peasant Agriculture and Economic Growth: The case of south-east Europe c.1870–1940 reinterpreted.* Working Papers 0028. European Historical Economics Society (EHES).

Leck, C., Evans, N. & Upton, D. (2014). Agriculture – who cares? An investigation of 'care farming' in the UK. *Journal of Rural Studies*, 34, 313–325.

Lobley, M. (1997). *Small scale family farming and the environment: The contribution of small farms to the management of 'conservation capital' in the British countryside.* Unpublished PhD thesis, University of London.

Lobley, M. (2000). Small-scale family farming and the stock of conservation capital in the British countryside. *Farm Management*, 10(10), 589–605.

Lobley, M., Errington, A., McGeorge, A., Millard, N. & Potter, C. (2002). *Implications of Changes in the Structure of Agricultural Businesses. Final report to Defra.* Plymouth: University of Plymouth.

Lobley, M. & Potter, C. (2004). Agricultural change and restructuring: Recent evidence from a survey of agricultural households in England. *Journal of Rural Studies*, 20(4), 499–510.

Lobley, M., Potter, C., Butler, A., Whitehead, I. & Millard, N. (2005). *The Wider Social Impacts of Changes in the Structure of Agricultural Businesses*. University of Exeter: Centre for Rural Policy Research.

Meert, H., Van Huylenbroeck, G., Vernimmen, T., Bourgeois, M. & van Hecke, E. (2005). Farm household survival strategies and diversification on marginal farms. *Journal of Rural Studies*, 21(1), 81–97.

Midmore, P. & Dirks, J. (2003). The development and use of 'rapid assessment methods' in ex-ante and ex-post evaluations of policy initiative in the rural economy. Paper presented to *Agricultural Economics Society Annual Conference.* Seale Hayne Campus, Plymouth University, Plymouth.

Milbourne, P., Mitra, B. & Winter, M. (2001). *Agriculture and Rural Society: Complementarities and conflicts between farming and incomers to the countryside in England and Wales*. Countryside and Community Research Unit Report to MAFF. Cheltenham: CCRI.

Murdoch, J. (1995). Governmentality and the politics of resistance in UK agriculture: The case of the Farmers' Union of Wales. *Sociologia Ruralis*, 35(2), 187–205.

National Assembly for Wales (2001). *Farming for the Future*. Cardiff: Government of the National Assembly for Wales.

Newby, H., Bell, C., Rose, D. & Saunders, P. (1978). *Property, Paternalism and Power: Class and control in rural England.* London: Hutchinson.

Newby, H., Rose, D., Saunders, P. & Bell, C. (1981). Farming for survival: The small farmer in the contemporary rural class structure. *In:* Bechhofer, F. & Elliott, B. (eds) *The Petite Bourgeoisie: Comparative studies of the uneasy stratum.* London: Palgrave Macmillan UK, 38–70.

Parry, J., Lindsey, R. & Taylor, R. (2005). *Farmers, Farm Workers and Work-related Stress: A report for the Health and Safety Executive*. London: Policy Studies Institute.

Potter, C. & Lobley, M. (1992). *Small Farming and the Environment: A report prepared for the Royal Society for the Protection of Birds*. Sandy: RSPB.

Potter, C. & Lobley, M. (1993). Helping small farms and keeping Europe beautiful: A critical review of the environmental case for supporting the small family farm. *Land Use Policy*, October, 267–279.

Pretty, J. (2013). *Agri-Culture: Reconnecting people, land and nature.* Abingdon: Earthscan.

Pritchard, B., Burch, D. & Lawrence, G. (2007). Neither 'family' nor 'corporate' farming: Australian tomato growers as farm family entrepreneurs. *Journal of Rural Studies*, 23(1), 75–87.

RDPE Network & National Care Farming Initiative (2010). *Support for Care Farming Through the Rural Development Programme for England: A review and look forward.* Bristol: Care Farming UK.

Rebanks, J. (2015). *The Shepherd's Life: A tale of the Lake District.* London: Penguin.

Redman, G. (ed.) (2015). *The John Nix 2016 Farm Management Pocketbook,* Melton Mowbray: Agro Business Consultants Ltd.

Reed, M., Lobley, M., Winter, M. & Chandler, J. (2002). *Family Farmers on the Edge: Adaptability and change in farm households.* University of Plymouth: Department of Land Use and Rural Management.

Saltmarsh, N., Meldrum, J. & Longhurst, N. (2011). *The Impact of Community Supported Agriculture.* Bristol: Soil Association.

Smaje, C. (2014). Kings and commoners: Agroecology meets consumer culture. *Journal of Consumer Culture*, 14, 365–383.

Smaje, C. (2015). The dearth of grass. *Land*, 18, 7–10.

Tudge, C. (2007). *Feeding People is Easy.* Italy: Pari.

Van der Ploeg, J. (2003). *The Virtual Farmer: Past, present and future of the Dutch peasantry.* Assen: Royal Van Gorcum.

Westbury, D. B., Park, J. R., Mauchline, A. L., Crane, R. T. & Mortimer, S. R. (2011). Assessing the environmental performance of English arable and livestock holdings using data from the Farm Accountancy Data Network (FADN). *Journal of Environmental Management*, 92(3), 902–909.

Wilson, M. (1996). Farmers and the market. *In:* Carruthers, S. & Miller, F. (eds) *Crisis on the Family Farm: Ethics or economics?* Reading: Centre for Agricultural Strategy, 236–240.

Wilson, P. (2016). *The Viability of the UK Small Farm: Analysis of Farm Business Survey 2014–15 data for England and Wales.* Specially commissioned report for Prince's Countryside Fund small farm research. University of Exeter: Centre for Rural Policy Research.

Winter, M. (1996). *The Working of an Affiliated Regional Advisory and Training Service (ARATS) for Environmental Land Management in England.* Consultancy report to FWAG and WWF.

Winter, M. & Lobley, M. (2016). *Is there a Future for the Small Family Farm in the UK?* Report to the Prince's Countryside Fund. London: Prince's Countryside Fund.

Zuckerman Committee (1961). *Scale of Enterprise in Farming: A report by the Natural Resources (Technical) Committee.* London: HMSO.

7 Farmers and social change

Stress, well-being and disconnections

Introduction

Chapter 3 described some of the more noticeable and quantifiable agricultural trends of recent years. While describing changes in farm size structures and land use, for instance, is an important part of the story of recent agricultural change, it is nevertheless only a partial picture of the changes experienced by farming families themselves. Once seen as important members of the community, both as employers and as participants in many of the key institutions of rural life (Ilbery, 1998; Murdoch and Marsden, 1994), as agriculture has become increasingly disconnected from both the surrounding community and the end customer, farmers have become increasingly socially and culturally isolated (Lobley et al., 2005; Pugh, 1996; Reed et al., 2002). This sense of isolation and disconnection has been compounded by a growing regulatory burden, financial volatility and the consequences of animal disease (Olff et al., 2005; Parry et al., 2005; Scott et al., 2004). In turn, this raises questions regarding the nature, extent and wider significance of the personal costs and social implications of agricultural change.

Against this background, this chapter considers some of the social and personal consequences of an agriculture that has developed to service globalised food systems and which operates in a countryside which itself has been socially, culturally and economically transformed, and which has arguably become increasingly disconnected from farming and farmers. Drawing on the 'stress iceberg' concept (see Lobley et al., 2004), it is argued that the well-publicised high rate of suicide among farmers actually masks the bulk of the issue, with many seeking professional or other help, simply suffering in silence or feeling generally unwanted and misunderstood. The causes of stress among farming families will be explored as will the notion that the disconnection between farmers, their local communities and their customers has left many feeling increasingly socially isolated, unwanted and unappreciated as farmers, and generally marginalised.

Rural social change

The British countryside has undergone significant social change over the post-war period. A number of factors including agricultural mechanisation (and the subsequent reduced demand for labour) and high rates of counter-urbanisation have led to agricultural production no longer occupying the place of central concern in rural communities. Where once it was the primary employer in rural areas, in 2016/17 agriculture, forestry and fishing only accounted for 7.5 per cent of rural employment and 1.3 per cent of total employment in England (Defra, 2018). Often characterised by rural geographers as the 'consumption countryside' (Marsden, 1999; Woods, 2011), rural space is arguably now dominated by the cultural services it provides in terms of aesthetic landscape and leisure and tourism opportunities. Ideas of a 'rural idyll' and improvements in transport and telecommunication links have also enhanced the countryside's attraction as a place to live. This has led to a recomposition in the population of rural areas, with subsequent friction between 'traditional' rural dwellers and 'incomers' being all too commonly cited (Bell, 1994; Cloke et al., 1995; Day, 1998; Mahon, 2007). Wider contestations over issues such as fox hunting (Milbourne, 2003), badger culling (Brumfiel, 2012), recreational access (Woods, 2003b), housing developments (Satsangi et al., 2010) and wind farms (Woods, 2003a) also evidence competing ideas about how the countryside should be governed. Hence, rural social change has inevitably altered the nature of farmers' interactions with their local community and may contribute to a sense of social, as well as geographical, isolation among farmers.

Alongside the declining prominence of farming in rural communities, other factors such as the rise of the supermarket (see Chapter 4) and a declining proportion of income spent on food have also been blamed for an increasing disconnection between the general public and agriculture. A stark indicator of this disconnection is a lack of awareness among children and young people about where food comes from: a survey by the British Nutrition Foundation in 2014 found that 18 per cent of primary school children thought bread came from animals and 24 per cent thought cheese came from plants (British Nutrition Foundation, 2014). Furthermore, according to a survey for the Prince's Countryside Fund, 12 per cent of 18 to 24 year olds have never seen a cow, 16 per cent have never visited a farm and 17 per cent have never visited the countryside at all (The Prince's Countryside Fund, 2017, p.6). A farmer interviewed for this book who hosts a large number of educational visits said, 'you get kids out here and they don't know things about where their food comes from and to tell them, you know, that you have to kill a pig before it can become pork, become sausages, some of them are quite blown away by that'. A lack of public understanding regarding food and farming not only has implications for health and education but, as we shall see, can also contribute to farmers feeling unacknowledged and undervalued for their role in the production of food (and maintenance of the countryside).

Stress and well-being in the farming community

Despite the considerable changes experienced by rural communities over the last few decades and the declining economic centrality of agriculture in all but the most remote rural locations, farm households are seen as an important social and cultural unit in the countryside (Gasson and Errington, 1993; Winter and Lobley, 2016). Relatively little, however, is known about their well-being in anything other than an economic sense. More recently it has been recognised that the changes in rural communities described above, and changes in agriculture such as the decline in farm labour (see Chapter 3), have contributed to a growing sense of disconnection and isolation among farmers, which may compound the difficulty of dealing with stress and which in turn has implications for the well-being of members of farm households.

Stress is all around us. It is a common aspect of life, be it urban or rural. According to the Samaritans *Stressed Out* survey (Samaritans, 2003), the majority of people experience some level of stress (82 per cent) and a significant minority (20 per cent) experience high levels of stress. Evidence suggests that work is a major source of stress. A 2013 survey for the mental health charity, Mind, found that 34 per cent of people said their work life was either very or quite stressful (Mind, 2013) and workplace stress can be an almost daily occurrence for individuals (Mazzola et al., 2011). Stress is not necessarily a bad thing, however. Indeed, a certain amount of stress is generally regarded as stimulating and life enhancing, although stress that leads to distress rather than a spur to activity or positive change can be hugely debilitating for individuals, families and, ultimately, communities. The exact effects of this distress vary between individuals, depending on their social, cultural and economic backgrounds, but can range from mild anxiety through to a potentially life threatening spiral of mental ill-health and suicidal ideation (thoughts of life not being worth living). The high rate of suicide among farmers is perhaps the most egregious indicator of mental distress within the farming population. Between 2011 and 2015 agricultural occupations had one of the highest elevated risks of suicide; second only to the building trade (Office for National Statistics, 2017b). Male suicides among skilled agricultural workers in this period had a standardised mortality ratio (SMR) of 169, and male suicides among elementary agricultural workers had a SMR of 191 (an SMR of 100 would indicate the same proportional level of suicides as the general population) (Office for National Statistics, 2017b). Similarly high rates were also recorded between 1993 and 2007 (Defra, 2007).

The seminal work on suicide among farmers in England and Wales by Hawton et al. (1998) explored the factors that led to suicide within a group of farmers between 1991–93 compared to a control group. Their findings revealed a complex interaction of contributory factors:

> Although problems at work and in family life were the most common difficulties, they appear less likely to be important in causing suicide than financial and legal problems and mental ill-health. The problems with work and family life seemed to raise the background level of stress,

> cutting off possibilities for support. Estrangement from the family may have been the last straw in a series of events.
>
> (Hawton et al., 1998, p.46)

Many of the 84 individuals studied had been diagnosed with a depressive disorder, many were being treated with anti-depressants and just over half had a history of psychiatric illness. In total, 58 out of the 80 people studied were assessed to have had a definite or probable mental disorder at the time of their death.

Although high rates of farming suicide are commonly found in many countries, it is not a universal phenomenon (Stark et al., 2006) and the exact nature of the link between suicide and the occupation of farming remains moot. On the other hand, what is without doubt is that, as well as the tragedy for the families and friends involved, the visible evidence of suicide represents just the 'tip' of a much larger mass of people suffering from stress and impaired personal well-being. Stress is commonly perceived to form part of a continuum of psychological distress where (to mix metaphors) suicide is seen as the tip of the 'stress iceberg'. Suicide is the part of the problem which is visible but under which lays a much greater body of depression, stress and anxiety. Thus, for every suicide there are a larger number of attempted suicides and it has been estimated that for every 100 people who are depressed and consulting a GP there are 400 people who are depressed and not consulting a GP (Jones et al., 1994). Therefore, for any clearly defined group within society (such as farmers), a high incidence of suicide indicates widespread but hidden stress and anxiety.

Although there may be a common-sense understanding of what stress is, there is no single medical condition known as 'stress'. It is a term used in conjunction with others (e.g. acute stress disorder, post-traumatic stress disorder) under the general umbrella of 'anxiety disorders'. Researchers have long faced the problem of finding an acceptable definition of stress. The difficulty is that stress is at once the cause and the effect … 'Stress, in addition to being itself, was also the cause of itself, and the result of itself' (Roberts, 1950, p.105). It is also an unavoidable consequence of life, promoting well-being up to a point, but resulting in ill-health beyond that point. This duality in the term has resulted in the stimulus now being called the 'stressor' while 'stress' describes the body's reaction. Others distinguish between stressors and strains. The latter being an individual's response to stress which can be physical, psychological or behavioural (Mazzola et al., 2011).

Stressors can be external (such as adverse physical conditions or stressful psychological environments) or internal (physical or psychological) and may be defined as short-term (acute) or long-term (chronic). Stress becomes chronic when the stressor/s are ongoing such as unremitting highly pressured work, long-term relationship problems, loneliness and persistent financial worries. Although precise definitions differ, a common element is that stress arises when there is an imbalance or lack of equilibrium between demand

and an individual's ability to meet that demand. Stress derives from an imbalance in the demands on a person and their ability to cope. In turn, an individual's coping ability will be influenced by their perceptions and values, as well as resources and skills. This implies a need to take into account the broad contexts of stress in people's lives rather than simply concentrating on narrower structures and stressors. The broader context is important because of its influence on the stressors to which people are exposed (Aneshensel, 1992 Cox, 1978; Sutherland and Cooper, 1990). This more sociologically informed approach emphasises the origins of a stressful life, implying the identification of causes, while clinical approaches are more concerned with illness as a consequence of stress.

Farming stress research

Some of the most important contributions to research on stress in farming arise from the work of McGregor and Deary (e.g. Deary et al., 1997; McGregor et al., 1995; 1996). In developing a methodological approach based on an earlier USA Farm Stress Survey (Eberhardt and Pooyan, 1990), McGregor, Deary and colleagues amassed considerable data on the causes of stress in farming and influenced subsequent researchers who have adopted similar methods (e.g. Boulanger et al., 1999; Firth et al., 2007). The approach involves completing a panel of questions relating to five stress 'domains' identified by Eberhardt and Pooyan: economics; geographical isolation; time pressures; climatic conditions; and hazardous working conditions. Employing this approach, a survey of farmers carried out at the Royal Highland Show in 1994 indicated that the highest ranking stressors were filling in government forms, adjusting to new government regulation and poor weather conditions. Machinery breakdown at difficult times, complying with environmental regulations, too much to do and too little time, and changes in the CAP scored slightly lower, inducing moderate rather than severe stress. Clearly, these results are highly sensitive to the time in which the research was conducted and it is likely that during periods of policy change, for instance CAP reform, this would have emerged as a more significant stressor. Similarly, financial worries did not rank as highly as other stressors, reflecting the fairly stable state of British agriculture in 1994.

A similar study carried out at the Royal Welsh Agricultural Show in 1998 (Boulanger et al., 1999) sought to replicate the research of McGregor et al. The research was undertaken in response to the high suicide rate for farmers in the county of Powys, which, based on the stress iceberg concept, was thought to obscure high levels of farming-related stress. The results of the survey conducted with farmers visiting the Royal Welsh Show indicate that government policy, finance, time pressure and the future of the family farm were top stressors. A substantive difference was found in comparison to the McGregor study in relation to financial pressure as a stressor. The Welsh study showed that significantly higher levels of stress were experienced by livestock farmers in contrast to the low levels reported by McGregor and colleagues, whose study had included a much higher proportion of arable farmers. Financial pressures, along with increased regulation, paperwork and workload (due to farm expansion and labour changes) were also found to be key stressors in a study by Parry et al. (2005) for the Health and Safety Executive. While this was true across farm types, the authors note that livestock farmers felt particularly under pressure due to financial challenges relating to seasonality, unpredictability, relatively low prices of meat, milk and wool, and the risk of animal disease.

A further study by McGregor et al. (1996) of arable and hill farmers on the east coast of Scotland aimed to assess their goals, objectives, attitudes and behaviour in certain areas of decision making. Farmers' decisions are influenced by a variety of factors, including stress and the ability to cope with it. Using the General Health Questionnaire, which measures psychological distress in a non-clinical situation, results showed that only 11 per cent scored highly (against an average of 20 per cent in the general population), but that these individuals also scored highly on the personality trait of neuroticism. This indicates that those farmers cope using their emotions rather than being task-oriented and blame themselves for problems arising, leading them to think they cannot cope. Based on these findings, McGregor et al. argue that, 'farmers falling into the high stress group will exhibit irrational decision making or make no significant decisions at all. This has obvious implications for the overall management and hence profitability of their business' (McGregor et al., 1996, p.234). In other words, understanding the well-being of farmers is important not just because of the implications for themselves and their families, but also because it impacts on the performance of agriculture as a business. Other aspects of farming performance such as animal welfare can also suffer as a result of farmer stress. A study in Denmark, for instance, found that financial difficulties, personal problems and poor mental health were associated with an increased risk of being convicted for animal neglect (Andrade and Anneberg, 2014).

One of the only other relatively large-scale studies of farming stress was undertaken in North Yorkshire on behalf of the Health and Safety Laboratory (Phelps, 2001). The survey was carried out in 2000 (prior to the major foot and mouth disease outbreak in 2001 but at a time when farmers were experiencing a considerable economic downturn) and sought to identify the level of stress among farmers, the contributory factors, and what coping and support mechanisms were being employed. The results of the analysis revealed that this group of farmers were experiencing high levels of anxiety, depression and stress. Half reported experiencing mild or severe anxiety and 24 per cent reported severe anxiety. A wide range of factors were identified as contributing to the level of mental ill-health reported by participants including financial problems, living/working alone, relationship problems/breakdown, and insufficient time to complete work/spend with family. Sixty-eight percent reported finding government policy 'very' or 'highly stressful', as was dealing with regulation and paperwork. Relatively few (13 per cent) reported that they found isolation to be 'very' or 'highly' stressful, although a larger proportion reported feeling isolated. Indeed, the Phelps study identified a statistically significant association between feelings of isolation, lone working, high levels of anxiety and depression:

> Just over a fifth of respondents stated that they felt isolated, which was also significantly associated with higher levels of anxiety and/or depression. As well as the physical isolation inherent in farming and the fact that the majority of farmers in this study worked alone most of the time, the finding that a number of respondents felt that not only the government was unsupportive, but that also the non-farming community did not understand or support farming may contribute to a feeling of isolation. The finding that the feeling of isolation is associated with high levels of anxiety and depression highlights the importance of social support networks as a coping mechanism.
>
> (Phelps, 2001, p.24)

Lone working, lacking a close confidant and limited social support networks have all been linked to farming suicide (Stark et al., 2006) and the factors that were identified by Phelps as potentially reducing the level of mental ill-health predominantly related to social support mechanisms; for example, having a confidant and seeing close friends and family regularly. Having someone to talk to is one of the most common coping strategies in response to stress across a wide range of occupations and workplace situations (Mazzolla et al., 2011). The absence of social support is compounded in the case of agriculture by gendered notions of farm work and masculine stoicism, which can deter farmers from asking for help (Alston, 2012; Bryant and Garnham, 2015).

The relationship between isolation and stress and well-being is complex, with different research methodologies producing different results. Generally, researchers employing the 'McGregor methodology' find that isolation is not a significant stressor but others identify isolation as an important stressor (see, for example, Peak District Rural Deprivation Forum, 2004). In a small-scale study of farming stress in Buckinghamshire, most farmers reported isolation as a cause of stress (Campbell, 2001). This was less a problem of physical isolation (although many farmers are still isolated from other people during their working day) and more to do with psychological isolation deriving from farmers' assumed poor standing in society and the changing nature of society, which was argued to increasingly not understand farming or farmers. While there is also evidence regarding the more positive attitudes of rural incomers and the wider public to farmers (see for example, Carruthers et al., 2013; Defra, 2010; Winter, 2003), this nevertheless suggests that the theme of 'reconnecting' widely promoted following the Curry Report on Farming and Food (Curry Report, 2002) should explicitly be broadened to include social reconnection as well as economic reconnection. Indeed, our study of the social implications of agricultural restructuring (Lobley et al., 2005) confirmed this sense of separation between farmers and the surrounding community and even pointed to a reluctance to identify as a farmer, as the following quotes from a discussion group illustrate:

> They're feeling persecuted, they're feeing vulnerable, they're feeling unwanted and they're feeling as if the whole world doesn't want farming. ...This alienation has been going on for quite a long time. ... Farmers actually don't feel that they want to stand up and show themselves in the community as being farmers.
>
> (Rural clergyman)

FARMER 1: I don't like telling people very much that I'm a farmer.
FARMER 2: No, you tend to shut up with that now. A few years ago ...
FARMER 1: You do. Twenty years ago you were a farmer and you were proud of it, and now, just like you say, you go there and you just keep your head down, you don't, well, unless you wanna annoy 'em.

Returning to the theme of isolation, another study (Raine, 1999) adopted a purposive sampling method to identify 20 farmers of different ages and operating farms of different sizes and types, and set out to discover through semi-structured interviews the causes and effects of stress on farmers. It focused on three main issues: farmers' perceptions of the stress involved in farming; the causes of stress; and the personal effects of stress. Most farmers interviewed considered farming to be stressful, and that it was becoming more so, with livestock farmers more stressed than arable (see also Boulanger et al., 1999). Not surprisingly, specific seasons were said to be more stressful than others (e.g. lambing, harvesting and planting). Lobley et al. (2005) also identified the impact of busy seasons on farmer stress and well-being:

> wife gets fed up of it really a bit, you know when we get to the end of the season. I get a bit fed up of not seeing all three of them, you know, the kids and the wife. I'm very tired at times. I make myself ill sometimes.
>
> (Farmer quoted in Lobley et al., 2005 p.35)

The impact of seasonality and long working hours aside, Raine (1999) identified three consistently cited stressors. Paperwork or bureaucracy was the most cited stressor regardless of farm type or size, while financial matters and the financial consequence of animal disease also emerged as important stressors. Isolation was a problem for those who worked alone, especially single hill farmers. Interestingly, those citing isolation as a stressor all reported to have strong family networks and good social lives but nevertheless felt isolated and stressed. This clearly suggests that, regardless of how well socially connected an individual may be, working alone and making day-to-day, on the job decisions alone contributes to elevated personal stress levels.

Ironically, as a result of farmers' own actions to restructure and streamline their businesses through labour shedding, farming has been a largely solitary occupation for some time, but as a consequence farmers often miss regular social contact during the working day:

> One thing I miss is dealing with the men I've had for 40 years. You know, I've known them man and boy as they say and I miss that. The day-to-day working with the men, with people.
>
> (Farmer quoted in Lobley et al., 2005 p.34)

> In farming, on a day-to-day basis you don't see many people. It's not like working in an office where you see the same crowd of people every day and you know them in and out and you see them five days a week when you are at work … you don't see people socially every day.
>
> (Farmer quoted in Lobley et al., 2005 p.34)

Unfortunately, the lack of social contact arising from lone working cannot always be resolved outside the working day, as the long hours associated with farming often leaves little time for relaxation and sociability. Thus, as we shall see through the empirical evidence presented below, there are a whole range of intertwining factors that put pressure on the well-being of the farming population, and which warrant a greater focus on the social health of the industry.

The well-being of farmers in the south-west of England

The research on stress and well-being discussed above informed our approach to the study of family farmers in the south-west of England via the 2010 and 2016 SW Farm Surveys. In considering the well-being of individuals taking part in the surveys, it is clear that the way farmers see themselves and their profession profoundly affects their sense of personal and collective self-esteem. A feeling that farming and farmers are misunderstood and undervalued by incomers to the rural community, by the urban majority and by government, was widely expressed. Personal well-being and self-worth are influenced by a complex range of factors, including not only personal economic success but also social and psychological factors that affect the subjectivity of an individual's opinion of his or her well-being. In particular, drawing on the model of Subjective Well-Being (SWB) advanced by Cummins (2005; 2010; Cummins et al., 2002) it can be argued that inputs from a farmer's environment, such as the type of farming system he or she manages, the level of family support, the intensity of familial and non-familial social networks, perceived attitudes of the public and officialdom towards farmers, will impact upon feelings of well-being but in a different way for every farmer depending on his or her underlying genetic disposition.

In order to explore how farmers see themselves, which in turn may be linked to other aspects of their subjective well-being, respondents were asked to supply a few words or phrases to describe 'what it is like to be a farmer in 2010/2016'.[1] The most common responses in 2010 included 'hard', 'paperwork' and 'too much', which together occurred over 680 times. Sometimes these words were combined, such as 'hard work with too much paperwork'. Similar descriptors were also common in 2016 – as visualised in Figure 7.1 – with a combined total of 589 occurrences of the words 'hard' (261 times), 'difficult' (107), 'challenging' (67), 'stressful' (52), 'tough' (45), depressing' (28) or 'frustrating' (29). Paperwork was still an issue for farmers in 2016, with 'paperwork', 'red tape' and 'regulations' together appearing 90 times, but there was also a notable emphasis on market pressures and unprofitability. The term 'little/low return' specifically occurred 41 times, but there were many more comments that in some way referred to financial challenges.

Across both surveys, for many there was a sense in which farming used to be hard work but also enjoyable work:

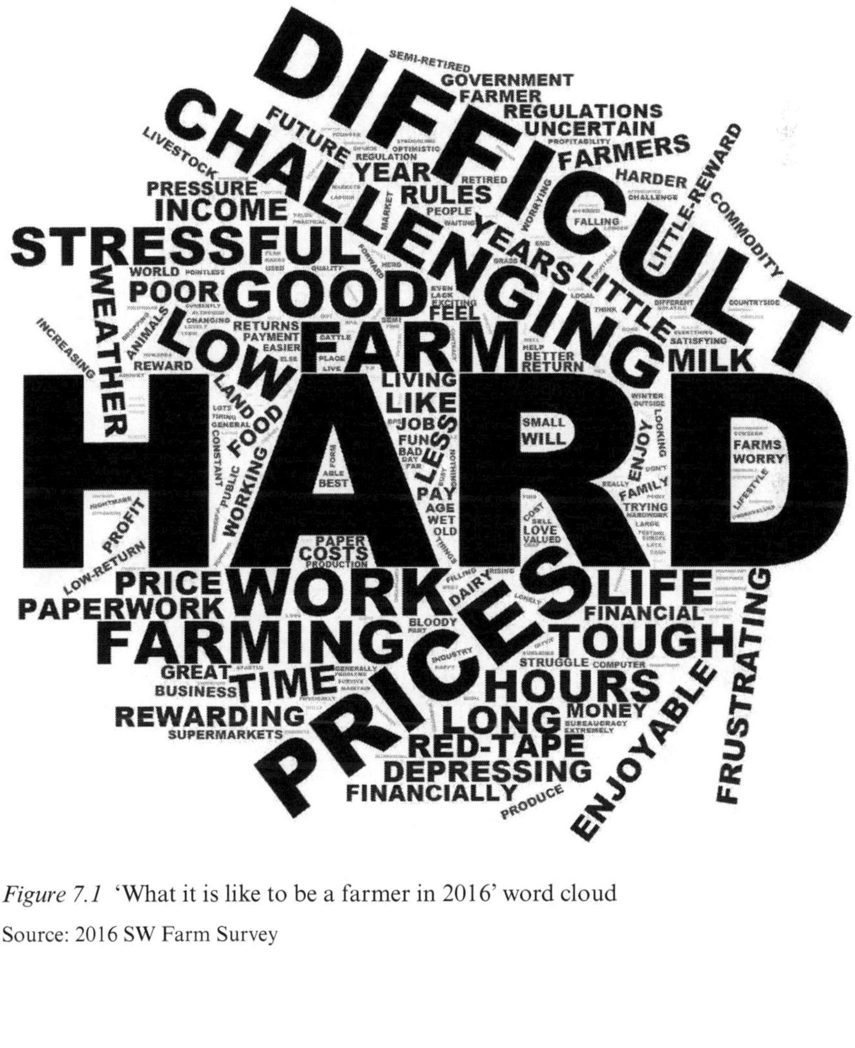

Figure 7.1 'What it is like to be a farmer in 2016' word cloud

Source: 2016 SW Farm Survey

> Twenty years ago farming was hard work and good fun. Now it is depressing, with too much red tape, forms for this, that and everything else.
>
> (Dairy farmer, 2010)

Now, all too frequently farming was seen as 'hard work with little reward' (Cattle/sheep farmer, 2010) coupled with a feeling that farmers are 'handcuffed by bureaucracy. Overwhelmed by officialdom. Stressed and overworked' (Cattle/sheep farmer, 2010). These were common terms in a discourse that positioned farmers as victims of the changing governance of agriculture in general and, in particular, the increase in 'form filling' and 'paperwork' which were often seen to get in the way of the 'real' job of farming: 'Hamstrung by red tape and form filling leaving less time for farming' (Mixed farmer, 2010); 'Spending 4–5 hours a day on paperwork. Animals are suffering, maintenance is being neglected' ('Other' farmer, 2010).

Low market prices were also leading farmers (particularly in the 2016 survey) to feel under incredible financial pressure, with little or no reward for their hard work:

> Depressing, soul-destroying. Poor commodity prices for all the hard work involved dents enthusiasm.
>
> (Dairy farmer, 2016)

> Extremely hard work to make a loss – not good for the soul. Unsustainable situation where farmers subsidise the world's cheap food. Bonkers.
>
> (Mixed farmer, 2016)

Market pressures were also related to strong feelings of being undervalued:

> I feel isolated and vastly underappreciated as a farmer. I used to feel proud that I was feeding the nation. Not one of us could survive without the food that we farmers grow, but no one seems to want to pay a decent amount for their food or even know or care where it comes from.
>
> (Cattle/sheep farmer, 2016)

> We are not appreciated. The milk price is so low another penny drop and we will be out of business. It's not a good time to be in farming.
>
> (Dairy farmer, 2016)

Interestingly, many farmers appear to be underestimating public opinion, as research shows attitudes towards farming have actually been consistently positive over the last decade (Carruthers et al., 2013). For instance, in 2012, 85 per cent of people in a YouGov-Cambridge survey agreed that 'farming is important for the UK economy as a whole', and 74 per cent agreed that 'farming plays an important role in protecting the environment in the UK' (University of Cambridge, 2012). Nevertheless, that this positivity is not being translated into farmer perceptions remains a cause for concern.

Other respondents spoke about being a farmer in terms of isolation, being trapped on the farm, being unwanted, and being tired, exhausted and 'generally misunderstood by non-farmers' (Mixed farmer, 2010). On-farm changes such as labour shedding and broader socio-economic change in rural communities left some feeling: 'lonely and isolated compared to 20–30 years ago' (Cattle/sheep farmer, 2010); 'like we are no longer wanted' (Cattle/sheep farmer, 2010); 'isolated, most neighbours sold up. Replaced by good lifers' (Cattle/sheep farmer, 2010). Similarly, another farmer told of the long, hard, tiring hours and that he did not 'feel understood anymore. More of the people we meet and come across haven't a clue about farming and what we do, even people in the village' (Dairy farmer, 2010).

Comments on the changing nature of rural life and the relative position of farmers were common:

> Because employment of local people by farms is much less than it was, because we can't afford decent wages, hence many foreigners who contribute nothing to village life. Also 30 years ago farmers held many prominent posts in rural communities but now they don't have the time.
>
> (Dairy farmer, 2010)

Narratives around the increasing presence of urban-rural migrants and a lack of understanding among the local community also appeared in response to the question 'Have your contacts with non-farmers changed over the last 5 years?' Many of the farmers who said their levels of contact had decreased felt that the change in the rural population is increasing their social isolation and eroding the traditional link between agriculture and rural communities:

> Small farms are getting less and farm conversions are bringing more town and city people into the country, many of which do not understand country life.
>
> (Cattle/sheep farmer, 2016)

> Many well-to-do incomers moving in. Most have no financial link to agriculture and minimal knowledge of same.
>
> (Arable farmer, 2016)

> Public come in the village, stay a short while and move again. Terrible, just to make money on a house!
>
> (Cattle/sheep farmer, 2016)

> Don't go anywhere much. Townies have no idea of country life.
>
> (Cattle/sheep farmer, 2016)

For some, this is contributing to a situation where 'it is us and them with the public – us and them!' (Cattle/sheep farmer, 2016) and a sense that 'no one seems to like farmers these days' (Arable farmer, 2016). At the same time there was a feeling that 'the general public don't care where their produce comes from as long as it's cheap' (Mixed farmer, 2010) and that 'food in this country is not valued' (farm type unknown, 2016). Some felt not only underappreciated but also victimised: 'No one is interested, no one wants us and we get the blame for everything – especially from the environmentalists' (Arable farmer, 2016). One farmer simply described feeling like 'a punch bag' (Mixed farmer, 2016). This sense of being unwanted was related to a belief that agriculture and rural communities had become disconnected: 'Forty years ago village people would have followed the farming year (asked about hay etc.). Now they tend to look at farming as an inconvenience to their enjoyment of rural living' (Horticulture farmer, 2010).

As we will see below, other farmers challenge these perceptions, arguing that agriculture and agricultural households remain the backbone of many rural communities and that farmers and the profession of agriculture is in a position to benefit economically and socially from an increased interest in the provenance of food generally and local food in particular. For now, though, it is important to note the potential cumulative impact on individuals who feel lonely and socially isolated, overly burdened by bureaucracy and unwanted.

Added to this are the effects of long hours of lone working which not only mean that being a farmer is 'physically and mentally exhausting' (Cattle/sheep farmer, 2010) but also that it is hard to take a break:

> Due to low income we cannot afford to employ anyone so the working hours are very long and can often be quite lonely not having time to get away and meet other people. There is no safety net for making mistakes.
> (Dairy farmer, 2010)

Another farmer commented that he had 'no quality of life. I work 7 days a week, often in excess of 16 hours a day. I have never had a holiday' (Cattle/sheep farmer, 2010). Another wrote of his constant battle against tiredness: 'If you could see me now as I'm trying to answer this survey – I can hardly stay conscious!!! I am always exhausted' (Mixed farmer, 2010). Clearly, long working hours, often in isolated and difficult conditions, can have implications for the well-being of farmers and may even contribute to the poor health and safety record of British agriculture. Agriculture has the highest rate of fatal injury among the main industrial sectors and one of the highest rates of major injury (Health and Safety Executive, 2016).

The impact of such tiredness and long working hours are not just confined personally to the responding farmer. Several also commented that as a consequence they did not spend sufficient time with their family and/or socialise outside of the farm. For instance:

> Farmers have to work extremely long hours to make their businesses viable. Not a good return for hours worked. This does make it hard to have enough time to spend with family and friends especially.
>
> (Dairy farmer, 2010)

> Stressful. Hard work. Less time for family life.
>
> (Mixed farmer, 2016)

These rather bleak and negative descriptions of being a farmer were not universally shared, however, with some farmers pointing to more positive developments such as becoming increasingly appreciated by the public, as the following quotes illustrate:

> At last it is getting better. At last the public realise we are going to be needed more and more for food production, landscape and environment, rights of way.
>
> ('Other' farmer, 2010)

> Enjoyable hard work. More appreciated than in recent past. Wonderful place to work.
>
> (Arable farmer, 2010)

> Exciting as there is a greater interest in food and where it comes from and food security is now on the agenda.
>
> (Mixed farmer, 2010)

> It's good that farmers and farming is back to where we were 50 years ago – respected and wanted!!!
>
> (Mixed farmer, 2016)

Some farmers also reported increased contact with non-farmers over the preceding five years and felt valued by the local community:

> Greater community awareness, support for local food, keenness to understand issue, environmentally positive.
>
> (Mixed farmer, 2016)

> Farming families are often the roots of the village and the one they come to in times of need.
>
> (Dairy farmer, 2016)

> We are the only farmers in our village now and live on the edge of a town. Public are very interested – ask about cows and calves and sheep and lambs in fields.
>
> (Cattle/sheep farmer, 2016)

It is notable, though, that such optimism regarding public interest in food and farming was far more frequent in 2010 than in 2016, when increasing market pressures were at the forefront of farmers' minds. Particularly low commodity prices at this time appeared to be overriding – or at least obscuring – the earlier enthusiasm about public appreciation, replacing it with a sense of being undervalued and taken for granted.

Not surprisingly, many farmers had mixed views and countered some of their own negative perceptions with prospects of improving incomes and increased demand for food:

> A challenging time with regulations but more rewarding as food is getting in short supply.
>
> (Dairy farmer, 2010)

> Challenging in terms of time. Exciting to build for the future. Rewarding in terms of quality of life.
>
> ('Other' farmer, 2010)

> Challenging, busy, stressful, rewarding, with a feeling of starting to be appreciated again like farmers used to be.
>
> (Arable farmer, 2010)

These examples indicate that self-esteem among some farmers is growing both as a consequence of a mounting interest in the provenance of food and because the market for the outputs of agriculture is seen to be set to expand significantly and rapidly. Again, such optimism about the future of farming was more apparent in 2010 than 2016, but in both surveys those with mixed views often felt that the challenges of farming were balanced by the type of life it affords, job satisfaction and the benefits of living and working in the countryside:

> It is long hours but it's still the best job – living the dream.
>
> (Mixed farmer, 2016)

> Challenging, but rewarding. A great way of life, but currently tough financially.
>
> (Dairy farmer, 2016)

> Fulfilling life, wonderful workplace, exciting potential. Currently challenging re: income and risk.
>
> (Arable farmer, 2016)

The 'great way of life' and 'wonderful workplace' described here point to the strong attachment that many farmers feel for their land and the farming way of life (Gray, 2000; Setten, 2005). This attachment is part of what makes farming more than simply an occupation for many individuals, driving them to continue agricultural production even in the face of low (or in some cases negative) profit margins. Some farmers also talked about finding their work 'on the whole rewarding' (Cattle/sheep farmer, 2016) and taking pride in what they do: '[I feel] fortunate to be doing a job of value to the country and the local community' (Cattle/sheep farmer, 2016). Such elements of job satisfaction are clearly beneficial for individuals' general well-being and can help to counter some of the stress that so many farmers reported. In some cases, however, job satisfaction is simply not felt strongly enough to override the pressures and challenges that farmers face.

On 'being a farmer', quality of life and well-being

In order to explore how respondents' experiences and outlooks around being a farmer (as described in response to the 'what is it like being farmer' question) are linked to other aspects of farm and personal/social life, each respondent was assigned to one of three categories[2] according to whether their overall portrayal of being a farmer was (i) exclusively or largely negative; (ii) exclusively or largely positive; or (iii) mixed. The normal caveats around the dangers and difficulties of labelling people according to their fluid and heterogeneous opinions apply but, for the sake of further analysis, we refer to each of these farmer groups as 'disenchanted, 'equivocal', and 'optimistic' respectively. Examples of how each of these farmer types described what 'being a farmer is like' are shown in Box 7.1.

Perhaps not surprisingly, in 2010 the majority of farmers (63.9 per cent) were disenchanted about being a farmer, although a fifth (19.4 per cent) were equivocal and a minority (16.7 per cent) were optimistic. By 2016 the picture had become even gloomier, with 75.2 per cent of farmers disenchanted and only 10.5 per cent optimistic (14.3 per cent were equivocal). In subsequent analysis this, admittedly rudimentary, typology was seen to be strongly associated with a range of variables associated with the farm, the farmer and their well-being and satisfaction. Interestingly, in 2010 optimistic farmers were significantly more likely to be operating an organic farm than disenchanted farmers (25 per cent of organic farmers expressed optimism compared to 14.8 per cent of conventional farmers and 16.7 per cent of the whole sample (χ^2=22.41, p<.001)). It would be dangerous to make bold statements about possible causality here but it is conceivable that for many farmers, having made a positive choice to convert to an organic system, they 'feel good' about being an organic farmer and being part of the broader movement of organic agriculture. One organic farmer, for instance, felt that 'being an organic grower selling fresh vegetables to a regular and loyal customer base at markets is very rewarding' (Horticulture farmer, 2010). Certainly, this supposition is consistent with the findings of a study by Mzoughi (2014), which found a positive and significant relationship between organic farming and high levels of life satisfaction among French farmers. However, in the 2016 sample the association between

Box 7.1 Views on being a farmer

Disenchanted
(63.9 per cent of farmers in 2010; 75.2 per cent of farmers in 2016)

These respondents described being a farmer in exclusively or largely negative terms. For example:

- 'Over worked, tired, stressed, constantly worried what the next disaster will be' (Mixed farmer, 2016).
- 'Hell' (Dairy farmer, 2010; Arable farmer, 2010; Cattle/sheep farmer 2010; 3x dairy farmers, 2016; Cattle/sheep farmer, 2016).
- 'Hopeless' (Mixed farmer, 2016; 2x dairy farmers 2016).

Optimistic
(16.7 per cent of farmers in 2010; 10.5 per cent of farmers in 2016)

These respondents described being a farmer in exclusively or largely positive terms. For example:

- 'Exciting. A privilege to be a food producer. A financial challenge. A pleasure to work in the country' (Cattle/sheep farmer, 2016).
- 'Absolutely fantastic, I love working with animals and people' (Cattle/sheep farmer, 2016).
- 'Excellent. How could you not like working in the great outdoors?' (Cattle/sheep farmer, 2016).

Equivocal
(19.4 per cent of farmers in 2010; 14.3 per cent of farmers in 2016)

These respondents expressed mixed or relatively neutral views on being a farmer. For example:

- 'Quality of life is good, but I work long hours for little return' (Cattle/sheep farmer, 2010).
- 'At times very rewarding, but in general very hard work, very stressful, lonely, frustrating and depressing. Over-regulated and full of deadlines' (Dairy farmer, 2016).
- 'Sometimes enjoyable. Sometimes frustrating with red tape & bureaucracy & poor weather' (Mixed farmer, 2016).

positivity and organic production had virtually disappeared, with only 12.6 per cent of organic farmers (compared to 10.3 per cent of conventional farmers and 10.5 per cent of the whole sample) optimistic about being a farmer. This decline in positivity was accompanied by numerous comments concerning low prices and financial difficulties, indicating that market pressures are reaching further across agriculture to the extent that the added value from organic food is no longer necessarily sufficient to provide protection from market pressures. The 'feel good' factor still exists for some, however, with the same organic horticulture farmer quoted above (who took part in both surveys) stating they were 'still proud to produce food in an environmentally responsible manner', though going on to add; 'Work/life balance? More life, please!' (Horticulture farmer, 2016).

In both 2010 and 2016, optimistic farmers appeared less likely to operate a dairy farm (in 2010 dairy farmers accounted for 27.9 per cent of the sample but only 12.3 per cent of optimistic farmers; in 2016 they represented 22.1 per cent of the sample but only 16.7 per cent of optimistic farmers[3]) and more likely to operate a cattle/sheep farm (in 2010 cattle/sheep farmers made up 40.6 per cent of the sample but 62.9 per cent of optimistic farmers; in 2016 they made up 38.7 per cent of the sample but 46.4 per cent optimistic farmers) (2010 $\chi^2=13.43$, $p=.037$; 2016 figures were not statistically significant). Dairy farmers have long felt that they are on a particularly arduous and relatively unrewarding treadmill due to low milk prices (which had just dropped sharply at the time of the 2016 survey) and this may account for the smaller concentration of dairy producers among optimistic farmers. The relative greater positivity among cattle/sheep farmers is more difficult to explain, particularly since this farm type has a poorer than average financial performance (see Chapter 3), but may be related to their age: in 2016, 40.8 per cent of cattle/sheep farmers (compared to 34.3 per cent of all farmers) and 42.5 per cent of optimistic farmers (compared to 24.8 per cent of all farmers included in the typology) were over the age of 65 (the relationship between optimism and being over 65 is significant; $\chi^2=24.71$, $p=.006$). Similar significant relationships were also observed in 2010. While farmer age is not necessarily an indicator of retirement status (as discussed in Chapter 3), it is likely that many of these older respondents were retired, semi-retired or winding down their involvement in the farm, and thus less marred by the immediate physical and economic challenges of the industry. This deduction is difficult to confirm without further research, however, and we would expect a combination of factors to underlie this observation.

There was also a very strong, significant association between attitudes towards being a farmer and a respondent's assessment of the current and future economic prospects for their business. As Figure 7.2 indicates, in 2010 optimistic farmers, and to a lesser extent equivocal farmers, were more likely to consider the current economic position of their business to be 'good' or 'excellent' compared to disenchanted farmers. Indeed, the latter are notable for their 'poor' assessment of the business both now and in the near future (see Figure 7.3). This is in contrast to the particularly sanguine outlook of optimistic farmers; 43.2 per cent of who view their economic prospects over the next 5 years as 'excellent' or 'good' compared to just 18.4 per cent of disenchanted farmers. The association between positivity on 'being a farmer'

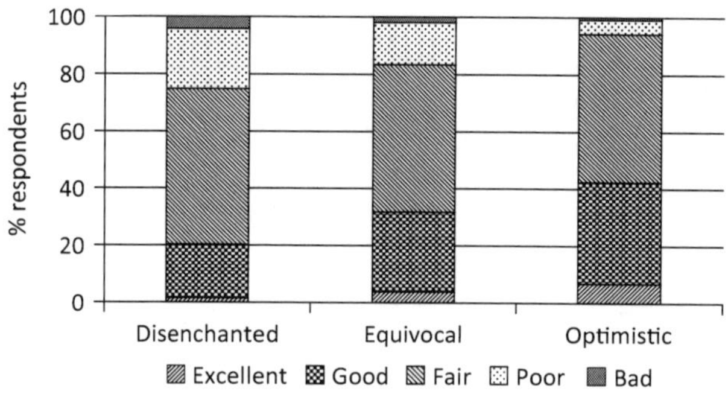

Figure 7.2 View on being a farmer by perceived current economic situation, 2010

Note: χ^2=77.57, p<.001.

Source: 2010 SW Farm Survey

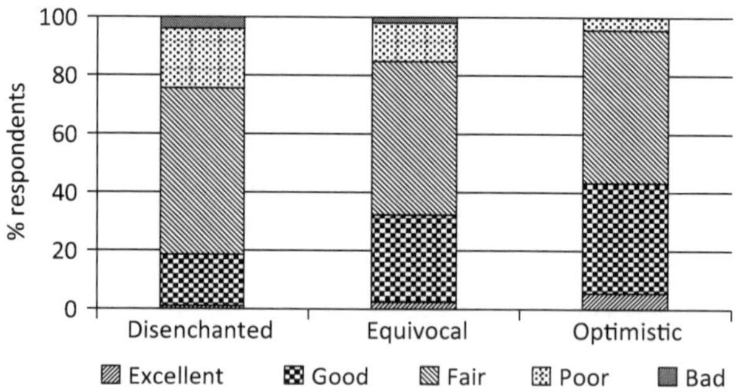

Figure 7.3 View on being a farmer by perceived economic prospects over the next five years, 2010

Note: χ^2=95.75, p<.001.

Source: 2010 SW Farm Survey

and perceived economic prospects was even more marked in 2016, with 79 per cent of optimistic farmers considering their economic prospects to be 'excellent' or 'good' compared to just 10.5 per cent of disenchanted farmers (Figure 7.4). In one sense it should not be surprising that a farmer's attitude about being a farmer is linked to their assessment of the performance of their business (as well as to feelings of being unwanted and misunderstood etc. as we have seen above). However, as we will see, the 'being a farmer' typology is also very strongly associated with a range of indicators of well-being and satisfaction with life, which suggest that optimistic and disenchanted farmers are on apparently divergent trajectories in terms of several aspects of their quality of life and subjective well-being.

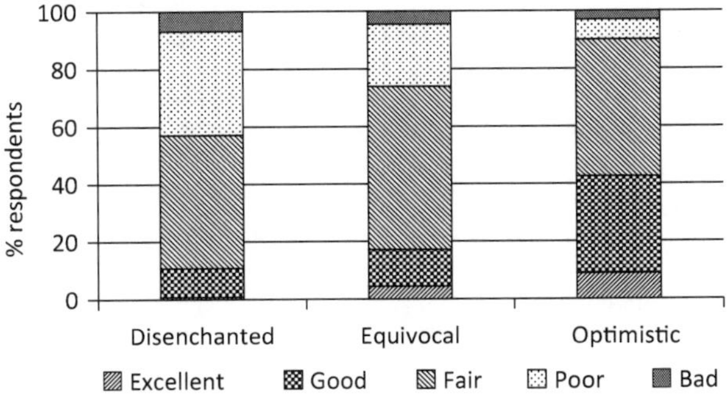

Figure 7.4 View on being a farmer by perceived economic prospects, 2016

Note: χ^2=101.04, p<.001.

Source: 2016 SW Farm Survey

Table 7.1 View on being a farmer by frequency meet with friends and relatives not connected to the business, 2010 (in %)

	On most days	*Once or twice a week*	*Once or twice a month*	*Less than once a month*	*Never*
Disenchanted	11.5	39.5	27.6	19.9	1.5
Equivocal	16.3	44.2	25.6	12.4	1.6
Optimistic	24.0	42.5	24.4	9.0	0.0

Note: χ^2=40.47, p<.001.

Source: 2010 SW Farm Survey

As an initial indicator of social support, farmers were asked about the frequency with which they meet with friends or relatives who are not directly connected to the business. As Table 7.1 indicates, in 2010[4] few disenchanted farmers met with friends and relatives on a daily basis and they were the most likely to report that they saw friends and relatives less than once a month. They were also the least likely to have seen their contacts with non-farmers increase over the preceding five years. Another indicator of the ability to get off the farm and mix with other people in potentially different situations is the frequency with which holidays are taken. Here it can be seen that in both 2010 and 2016 close to a quarter of disenchanted farmers claimed to never take a holiday (see Tables 7.2 and 7.3). We can only guess at the impact of not being able to take a break from doing something that one has a negative opinion of; of being 'trapped' in the workplace of a job that many of these farmers feel is unappreciated and beset by unnecessary 'interference' from government and which, as we have seen, many find mentally and physically

Table 7.2 View on being a farmer by frequency take holiday, 2010 (in %)

	More than once a year	*Once a year*	*Less often than once a year*	*Never*
Disenchanted	15.1	31.6	30.1	23.2
Equivocal	20.9	28.3	30.6	20.2
Optimistic	33.2	34.1	23.6	9.1

Note: $\chi^2=51.09$, $p<.001$.

Source: 2010 SW Farm Survey

Table 7.3 View on being a farmer by frequency take holiday, 2016 (in %)

	More than once a year	*Once a year*	*Less often than once a year*	*Never*
Disenchanted	22.1	28.6	28.0	21.4
Equivocal	29.5	34.5	24.5	11.5
Optimistic	44.6	25.7	13.9	15.8

Note: $\chi^2=34.02$, $p<.001$.

Source: 2016 SW Farm Survey

exhausting. In contrast, optimistic farmers are the least likely to never take a holiday and are significantly more likely to take more than one holiday a year – something much easier to achieve for those not tied to dairying or livestock production. On their own, however, these statistics mean little. Perhaps these farmers don't want to interact with others, other than during the course of their day-to-day business. Perhaps the famous farming stoicism means that they are content and self-contained. That is undoubtedly the case for some, though probably a small minority. As one farmer put it; 'I don't need to connect much with people outside of the industry"'(Cattle/sheep farmer, 2010). Another expressed their views in particularly vibrant terms: 'Can't stick non-country, non-farming people. Clueless, nasty, mean, judgemental, jealous money grabbers' (Dairy farmer, 2010). Such fervently expressed negative views were very rare, although farmers disenchanted with their farming occupation were anything but rare and, as we will see, this same group of farmers tend to be dissatisfied with their leisure time and their social life.

As we have seen, many farmers are working long hours and regularly suffer from tiredness. Table 7.4 reports on farmer satisfaction with the amount of leisure time at their disposal in 2010.[5] Those with an optimistic outlook on being a farmer appear to be particularly satisfied with the amount of leisure time in their lives, with 77.8 per cent reporting that they were *satisfied*, *more*

Table 7.4 View on being a farmer by satisfaction with leisure time, 2010 (in %)

	Not at all satisfied	*Less than satisfied*	*Satisfied*	*More than satisfied*	*Completely satisfied*
Disenchanted	16.5	43.5	36.0	3.0	1.1
Equivocal	7.8	40.7	45.0	5.4	1.2
Optimistic	3.6	18.6	58.6	11.4	7.7

Note: $\chi^2=141.78, p<.001$.

Source: 2010 SW Farm Survey

Table 7.5 View on being a farmer by satisfaction with social life, 2010 (in %)

	Not at all satisfied	*Less than satisfied*	*Satisfied*	*More than satisfied*	*Completely satisfied*
Disenchanted	8.2	27.9	52.0	7.6	4.2
Equivocal	2.0	22.3	55.1	14.1	6.6
Optimistic	0.5	6.8	59.3	22.6	10.9

Note: $\chi^2=112.24, p<.001$.

Source: 2010 SW Farm Survey

than satisfied or *completely satisfied* with the current situation. This can again be contrasted with the lower levels of satisfaction and quite marked dissatisfaction of farmers expressing disenchantment about being a farmer. Just over 40 per cent report being *satisfied*, *more than satisfied* or *completely satisfied*, while a significant minority of 16.5 per cent (139 individuals) report being *not satisfied at all*. Again, those expressing equivocal attitudes about being a farmer occupy the middle ground. Table 7.5 reports similar data (again for 2010) but this time for a farmer's social life.[6] The pattern of results is largely consistent with what we have seen so far. Just over 36 per cent of disenchanted farmers report being *less than satisfied* or *not satisfied at all* with their social life compared to just 7.2 per cent of optimistic farmers. In contrast, 33.5 per cent of optimistic farmers report being *more than satisfied* or *completely satisfied* with their social life compared to 11.8 per cent of disenchanted farmers.

These data point to a number of strong statistical associations between attitude towards being a farmer and a range of indicators that would generally be considered to reflect quality of life. It must be noted that, in themselves, these statistical associations tell us nothing about causality. Indeed, it is likely that there is a complex interaction between a range of variables, low levels of satisfaction and disenchantment about being a farmer. Moreover, in addition to indicators of individual aspects of life such a leisure time, we can see from Tables 7.6 and 7.7 that disenchanted farmers are particularly

Table 7.6 View on being a farmer by satisfaction with life in general, 2010 (in %)

	Not at all satisfied	*Less than satisfied*	*Satisfied*	*More than satisfied*	*Completely satisfied*
Disenchanted	5.7	29.1	53.3	9.1	2.7
Equivocal	1.6	16.3	55.8	22.9	3.5
Optimistic	1.4	4.5	45.5	36.4	12.3

Note: $\chi^2=191.60$, $p<.001$.

Source: 2010 SW Farm Survey

Table 7.7 View on being a farmer by satisfaction with life in general, 2016 (in %)

	Not at all satisfied	*Less than satisfied*	*Satisfied*	*More than satisfied*	*Completely satisfied*
Disenchanted	4.4	26.1	52.5	13.5	3.5
Equivocal	0.0	10.8	57.6	26.6	5.0
Optimistic	1.0	1.0	34.7	44.6	18.8

Note: $\chi^2=140.69$, $p<.001$.

Source: 2016 SW Farm Survey

dissatisfied with their life in general. This also applies to equivocal farmers but to a lesser extent. Some 34.8 per cent of disenchanted farmers in 2010, and 30.5 per cent in 2016, reported that they are dissatisfied with life in general, compared to 6.1 per cent in 2010 and only 2 per cent in 2016 of optimistic farmers. These findings can be compared to figures from the Annual Population Survey (APS), which indicate that in 2015/16 just 4.7 per cent of people in England (3.7 per cent in the South West) reported low levels of life satisfaction and 81.2 per cent (83.4 per cent in the South West) reported high or very high levels of life satisfaction (Office for National Statistics, 2017a). Although not directly comparable (i.e. the APS employed a 10-point scale whereas our study used a 5-point scale) this data nevertheless suggests that disenchanted farmers are much more dissatisfied with their lives than the general population and that optimistic farmers are somewhat more likely to be satisfied with life. Of note, while life satisfaction appears to have declined among farmers between 2010 and 2016, life satisfaction among the general population has increased over a similar time period (mean scores for life satisfaction in England rose from 7.4 in 2011/12 to 7.6 in 2015/16, and from 7.5 to 7.7 in the South West), reflecting the unique situation of farmers in terms of their lifestyle, occupation and sensitivity to the financial climate of their industry.

Finally, evidence from our surveys suggests that, for a significant proportion of disenchanted farmers, the situation is deteriorating. In 2016, 37 per cent reported being less satisfied with life than they were 12 months previously (compared to 20.3 per cent and 5 per cent respectively of equivocal and optimistic farmers). On the other hand, 7.5 per cent reported that they were now more satisfied with life than they were 12 months previously, although this compares to 14.5 per cent and 23.8 per cent respectively of equivocal and optimistic farmers. These findings were statistically significant (χ^2=*66.20, p<.001*). On a positive note, this confirms research that indicates that after a setback life satisfaction levels can (but do not always, depending on the strength of the negative experience) return to the equilibrium level of general satisfaction with life (Clark and Georgellis, 2013; Cummins, 2010; 2003; Cummins et al., 2002). Less positively, however, it points to a significant minority of farmers that may be entering a downward spiral of increasing dissatisfaction with life alongside a feeling of being unwanted, misunderstood, overloaded by paperwork and overworked but underpaid. That farmers often feel underappreciated by the local community and wider public (as we have already seen) also has an impact on their overall opinion of what it is like to be a farmer and, therefore, on their general sense of well-being. In 2016, we observed a relationship between opinions of being a farmer and the extent to which farmers agreed with the statement 'I feel valued'[7] (see Table 7.8). Optimistic farmers were significantly more likely to agree with the statement than disenchanted farmers (39.8 per cent of optimistic farmers agreed or strongly agreed, compared to just 18.7 per cent of disenchanted farmers). Although not directly comparable, these figures are markedly lower than the almost three-quarters of the UK population who report feeling that they are 'treated with respect' (Huppert et al., 2009).

Respondents who were optimistic about being a farmer were also significantly more likely to agree with the statements 'I feel optimistic', 'I feel relaxed' and 'I feel what I do is worthwhile' (see Tables 7.9 to 7.11). Notably, as with levels of life satisfaction, farmers generally reported lower levels of feeling what they do is worthwhile (see Table 7.9) than the general population. Where 56.7 per cent of all farmers agreed or strongly agreed with this statement and

Table 7.8 Levels of agreement with the statement 'I feel valued', 2016 (in %)

	Strongly disagree	*Disagree*	*Neither disagree nor agree*	*Agree*	*Strongly agree*
Disenchanted	26.0	27.9	27.3	13.2	5.5
Equivocal	10.6	27.3	36.4	22.0	3.8
Optimistic	5.1	17.3	37.8	28.6	11.2
All farmers	**21.5**	**26.7**	**29.8**	**16.2**	**5.9**

Note: χ^2=*56.48, p<.001*.

Source: 2016 SW Farm Survey

Table 7.9 Levels of agreement with the statement 'I feel optimistic', 2016 (in %)

	Strongly disagree	*Disagree*	*Neither disagree nor agree*	*Agree*	*Strongly agree*
Disenchanted	19.4	24.8	35.2	15.6	5.0
Equivocal	4.5	18.8	37.6	34.6	4.5
Optimistic	3.1	4.1	34.7	34.7	23.5
All farmers	**15.5**	**21.7**	**35.5**	**20.4**	**6.9**

Note: $\chi^2=119.24, p<.001$.

Source: 2016 SW Farm Survey

Table 7.10 Levels of agreement with the statement 'I feel relaxed', 2016 (in %)

	Strongly disagree	*Disagree*	*Neither disagree nor agree*	*Agree*	*Strongly agree*
Disenchanted	18.6	28.4	31.7	15.5	5.8
Equivocal	5.3	22.1	42.7	26.0	3.8
Optimistic	5.1	10.2	27.6	41.8	15.3
All farmers	**15.2**	**25.5**	**32.9**	**19.8**	**6.6**

Note: $\chi^2=83.12, p<.001$.

Source: 2016 SW Farm Survey

Table 7.11 Levels of agreement with the statement 'I feel what I do is worthwhile', 2016 (in %)

	Strongly disagree	*Disagree*	*Neither disagree nor agree*	*Agree*	*Strongly agree*
Disenchanted	10.0	14.6	24.0	29.4	22.0
Equivocal	1.5	11.2	21.6	43.3	22.4
Optimistic	4.0	3.0	11.1	39.4	42.4
All farmers	**8.1**	**12.9**	**22.3**	**32.5**	**24.2**

Note: $\chi^2=51.53, p<.001$.

Source: 2016 SW Farm Survey

8.1 per cent strongly disagreed, 83.6 per cent of people in England in 2015/16 (85.3 per cent in the South West) indicated high or very high levels of feeling what they do is worthwhile and just 3.6 per cent (3.3 per cent in the South West) indicated low levels of feeling what they do is worthwhile (Office for National Statistics, 2017a).

Many of the narratives around 'what it is like being a farmer' that emerged from the 2010 and 2016 SW Farm Surveys were also reflected in the national research conducted for SIP in 2015, where the same question was asked (see Morris et al., 2016). As in the South West, land managers in the SIP research were predominantly concerned about paperwork and financial challenges but also cited benefits arising from job satisfaction and the farming 'way of life'. However, overall levels of positivity/negativity appear to be more balanced in the SIP research than the principally negative opinions of the South West farmers, with 37 per cent holding exclusively or largely positive opinions, 37 per cent exclusively or largely negative opinions and 25 per cent mixed or neutral opinions. This more balanced picture may reflect the broader spectrum of farming conditions at the national scale, as the SIP research encompassed more large/very-large/ultra-large farms and more arable systems than the South West research.[8] The poorer financial performance of South West farms compared to the national average is likely to contribute to this outcome, particularly as the farmers of smaller, livestock and dairy units that characterise agriculture in the South West may be more vulnerable to having their morale eroded by factors such as low prices and market volatility. Indeed, dairy and mixed farmers in the 2016 SW Farm Survey (which together made up 47.6 per cent of the sample) were both more likely than expected to express negative opinions on being a farmer and both these farm types made an average loss in terms of return on capital employed (ROCE) in 2015/16 (Defra, 2017). Contradictorily, as noted earlier, cattle/sheep farmers were the most positive about being a farmer despite livestock farms (both lowland and LFA) having the worst levels of ROCE of all farm types in 2015/16. This anomaly might be explained, however, by the observation that – either due to higher than average performance within the survey sample or simply lower expectations – cattle/sheep farmers' perceptions of their economic prospects in the 2016 SW Farm Survey were fairly similar to the average across all farm types, with 48.1 per cent of these farmers considering their prospects to be 'fair'.

(Re)connections

While the impacts of time pressures faced by farmers, broken social connections and feelings of mistrust are largely experienced at the farm and farm household scale, they are not confined to farmers and their families. Evidence suggests that some of the traditional institutions of rural and farming life such as the village pub, church and market are in decline, partly as a consequence of the changing composition of rural populations but also because of 'competing time pressures on farmers' (Parry et al., 2005, p.65). Quite simply, a survival strategy of working harder and longer takes its toll. Appleby (2004) contends that declining social networks are due largely to farmers' lack of time for social contact, coupled with the erosion of natural links between farms, farmers and local communities.

In contrast to the somewhat gloomy picture to emerge from much of this chapter, the discussion now turns to how different ways of being a farmer can be associated with improved personal well-being and (re)connections with communities and customers. We present below some illustrative examples which suggest that for some farmers and their families, stepping off the treadmill can yield a well-being dividend. The concept of the agricultural treadmill was developed by Cochrane (1958; 1979) to describe the process whereby farmers find themselves having to 'run' simply to survive. As we have seen, farmers often use language redolent of the treadmill metaphor when describing the ever-increasing workload just to stand still. The farming family as a whole often take the strain as the principal farmers spend ever longer hours working in an attempt to help the business survive, which inevitably leaves less time for their children or social and leisure activities. As we have seen, this can be particularly marked for livestock producers and dairy farmers, where the price of a heavy commitment of personal time may be paid in feelings of dissatisfaction with life. What then is the impact of stepping off the treadmill, of making a decision to do things differently, to find another way of being a farmer?

One farmer interviewed by Lobley et al. (2005) restructured his business and began to disengage from active farming because of the impact it was having on his family relationships and the experience of friends who had been in similar situations:

> We used to do contract farming for other people and we had four full-time people and part-time. We used to do a lot of work for other people on a contract basis. I cut all that out because my present wife said, 'well I don't want to know you if you are working every weekend', so that was a conscious decision as well. Because I've had three of my friends get divorced in the last two years because they've actually got busy and they're working every weekend. You know, on a tractor, or feeding pigs or doing, you know, doing something every weekend and therefore, they never saw their children.

The same farmer went on to describe the well-being dividend resulting from his decision to restructure the business:

> You can get up on a Sunday morning and read the Sunday paper without worrying about feeding livestock. And you can go and see your friends for the weekend … You can have holidays when you want them rather than being dictated by the weather or your business.

Another farmer dramatically restructured his business. He decided to let his land out while earning a living as an agricultural and agri-environmental contractor. The result was quite simply that 'well, the kids have got their dad back'. He went on to discuss the impact on his own personal well-being:

> I can go out on a Friday night and not worry about getting home because I've got to get up for milking at 5.30. I can lay in bed on a Saturday morning. I can go out with my family on a Sunday. I can watch football

> and cricket. I can have a day off when I want; I haven't got to get back for milking. Never have to get up to calve a cow in the middle of the night, like. That's brilliant. I don't stink of sour milk and cow shit. I had five days holiday in 15 years I think. I had five days off. Other than that I was here milking and feeding.

In both these cases the interviewees identified positive changes in their family relationships, social networks and leisure time. Admittedly, both had radically restructured their business and had become more disengaged from agriculture. Others restructure and realign their business. The operator of one farm described a number of changes to his business over the years stemming originally from recognising the futility of attempting to intensively farm land that was often of poor quality and topographically challenging, requiring double-wheeled tractors to help prevent them from tipping over on steep banks. Following a decision in 2006 to cease organic milk production, the farm now operates a range of enterprises including beef, pigs, vegetables, a meat box scheme, educational events and a range of diversified activities.

Speaking about the decision to leave dairying, the farmer said:

> I do miss the sense of satisfaction of seeing a herd of healthy organic dairy cows, it's quite satisfying … But otherwise no. I don't miss, almost two weeks after I gave up, after we sold them, I realised that 'oh my God, this is fantastic, I can lie in in the mornings. 7 o'clock start instead of a 5 o'clock start'. And didn't have so many animals to look after, that was another thing, the numbers, the health of each one.

The changes on the farm have been progressive over a number of years and:

> it's been a bit, not all brilliant. Because you see there's a lot more people here and just my time is often taken up with managing people and meeting people's needs, and it's a bit frustrating sometimes. It can be a bit too much.

Clearly, increasing contact with customers and other visitors is not for everyone but the farmer described how 'I also love to have people here as well. It's getting a balance'. He went on to explain that:

> You get more people from the village coming up. Stronger connections with local people. Definitely. Big impact. Because before I was selling a commodity product, milk … but I kind of let go of all that now and we've got the vegetables and we've got, selling the pumpkins, people come to the farm to buy them and we've got the meat box. It's much more interactive with the local community.

In reflecting on how the farm had changed and the impact of this, the farmer said:

> No, looking back on it it's been an amazing journey and (that sounds a bit cliché or something) but no I wouldn't change anything and I guess, yeah, the people I've met over the years and the places I've been have allowed me to open up a bit … I always used to hate the culture, [which] is very much the Englishman and his castle, it's a bit about land we tend to want to own. And so now it's very much about sort of 'get on my land' rather than 'get off my land'. And it's … a big culture change for me to work with people with those qualities and it's quite hard but once you really take it on you have this wonderful experience. And so that has been, the part about embracing as many people as possible, but as I said before, you can't do everything. You can only have so many people here. But yeah, that's what it's been about for me anyway.

This famer has actively sought out a different way of being a farmer. Much of what he does will be familiar to any farmer, it's just, as he explains 'a bit different':

> I've had three options in the last 20/30 years. Either to get bigger, to give up or to diversify. So I've chosen the last one. Sort of embraced that one. So that's what it's been about for me, yeah, willing to diversify. But you've still got to do the normal things that every other farmer has to do, every other businessman has to do. It's a business, that has to run as a business so that's part of the farming and very much so … I mean you've still got to get up in the morning to organise the day. You know, still not that much difference in terms of [the] structure of running a dairy herd and producing milk. All the animals have to be looked after, cared for, fed and I have to organise that all everyday still. So there's still basic things that a farmer would do, but I suppose it's just a bit different.

Conclusions

With our concern for the 'social' and the 'personal', this chapter has revealed widespread contemporary evidence of previously reported stressors such as paperwork, social isolation and the changing position of farmers in society. There is also evidence from the SW Farm Survey that farmers are less likely than the general population to feel that what they do is worthwhile. As ever, the situation is complex and such is the heterogeneity of the farming population that two neighbouring famers could have vastly different perceptions of their role in society, how valued they feel and the extent to which they feel isolated.

Certainly, we have identified a large core group of farmers who are dissatisfied with many aspects of their lives, including that of being a farmer. Although we cannot ascertain evidence of suicide and clinical depression among our sample, the bulk of these dissatisfied/unhappy farmers lends weight to the stress iceberg hypothesis. A large amount of human misery within the farming population is hidden beneath the surface. The social, as opposed to physical, isolation experienced by many of these farmers means they may lack the important personal and social buffers necessary to maintain their esteem and levels of life satisfaction/well-being. In turn, this suggests the continuing need for the role played by the farming help charities. Indeed, it may be argued that these support groups will play an even more important role in the short to medium term as farmers, depressed and unhappy or otherwise, come to terms with a post-Brexit future. At the time of writing the shape of post-Brexit domestic agricultural policy and trade policy remains unknown but, for those already close to the edge, a future of reduced domestic support and exposure to world markets may be an unwelcome development.

Those with a more positive attitude about being farmers score much better across a range of well-being/life satisfaction indicators. It may be a step too far to call them unrelentingly cheerful, but they certainly appear happier than their colleagues who are disenchanted with being a farmer. Those who were more equivocal about their role as a farmer occupy the middle ground. We are not able to attribute causality to factors that determine membership of any of these three groups but in addition to underlying personality traits, personal and family situations there is also a complex interaction between farm type, age and other characteristics. What is clearer, however, is that the interaction between views on being a farmer and a range of attitudinal and subjective well-being indicators points to the potential for disenchanted farmers to enter a downward spiral of feeling bad about themselves, their work and lives; feeling unwanted, misunderstood and unsupported. Not only is this a challenging and difficult position to move forward from, but it may mean that what optimistic farmers see as an opportunity, disenchanted farmers perceive as yet another threat to the existence and continuity of their family business.

We have also seen that there are different ways of being a farmer. That for some, stepping off the agricultural treadmill provides a new lease of life; a personal and familial well-being dividend. This is not to suggest that farmers who are mainstream commodity producers working long hours can't be happy and have fulfilling lives, but that there are different ways of farming, and ways of reconnecting with neglected family and the local community. In a post-Brexit world, these different ways of farming, of being a farmer, may become more distinct and more polarised. There is arguably a case for government support for those who decide to move their business onto a different pathway. There will most certainly be an even greater need for the work of the UK's Farming Help Charities[9] as farming families adjust to the new reality and as some, by choice or otherwise, leave the industry and their home.

Notes

1 This question was not asked in 2006.

2 A total of 1,332 farmers in 2010, and 961 farmers in 2016, responded to the question 'what is it like being a farmer in 2010' in a manner that could be assigned to our three categories.

3 In order to obtain meaningful results from the statistical test, horticulture, pig, poultry and 'other' farms have been excluded from this analysis, as each accounted for less than 2 per cent of both the 2010 and 2016 sample (a combined total of 5.3 per cent in 2010 and 7.4 per cent in 2016).
4 This question was not asked in 2016.
5 This question was not asked in 2016.
6 This question was not asked in 2016.
7 This question was not asked in 2010.
8 Mean farm size in the SIP Baseline Survey was 273 ha, compared to 139 ha in the 2010 SW Farm Survey and 134 ha in the 2016 SW Farm Survey. Some 23% per cent of the SIP sample were arable farms (either cereals or general cropping), compared to 7.3 per cent and 7.7 per cent of the 2010 and 2016 SW Farm Survey samples respectively.
9 Together the UK's Farming Help Charities (The Addington Fund, The Farming Community Network and The Royal Agricultural Benevolent Institution) work in different ways to provide various forms of support, from homes for farming families who have to leave their farm and by doing so will lose their home, help to retired and working farming people in financial difficulty and confidential support for farming, business, health and family matters. www.farminghelp.co.uk.

References

Alston, M. (2012). Rural male suicide in Australia. *Social Science & Medicine*, 74(4), 515–522.

Andrade, S. B. & Anneberg, I. (2014). Farmers under pressure: Analysis of the social conditions of cases of animal neglect. *Journal of Agricultural and Environmental Ethics*, 27(1), 103–126.

Aneshensel, C. S. (1992). Social stress: Theory and research. *Annual Review of Sociology*, 18(1), 15–38.

Appleby, M. (2004). *Norfolk Arable Land Management Initiative (NALMI). Final project report*. London: Countryside Agency.

Bell, M. (1994). *Childerley: Nature and morality in a country village*. Chicago: University of Chicago Press.

Boulanger, S., Gilman, A., Deaville, J. & Pollock, L. (1999). *Farmers' Stress Survey. A questionnaire carried out at the Royal Welsh Show into stress factors experienced by farmers*. Institute of Rural Health.

British Nutrition Foundation (2014). *National Pupil Survey 2014: UK survey results*. London: British Nutrition Foundation.

Brumfiel, G. (2012). Badger battle erupts in England. *Nature*, 490(7420), 317.

Bryant, L. & Garnham, B. (2015). The fallen hero: Masculinity, shame and farmer suicide in Australia. *Gender, Place & Culture*, 22(1), 67–82.

Campbell, D. (2001). Stress in Buckinghamshire Farming. Quainton: Bucks Rural Health Group.

Carruthers, P., Winter, M. & Evans, N. (2013). *Farming's value to society: Realising the opportunity*. Report commissioned by the Oxford Farming Conference. Oxford: The Oxford Farming Conference.

Clark, A. E. & Georgellis, Y. (2013). Back to baseline in Britain: Adaptation in the British Household Panel Survey. *Economica*, 80(319), 496–512.

Cloke, P., Goodwin, M. & Milbourne, P. (1995). 'There's so many strangers in the village now': Marginalization and change in 1990s Welsh rural life-styles'. *Contemporary Wales*, 8, 47–74.

Cochrane, W. (1958). *Farm Prices, Myth and Reality.* Minneapolis: University of Minnesota Press.

Cochrane, W. (1979). *The Development of American Agriculture: A historical analysis.* Minneapolis: University of Minnesota Press.

Cox, T. (1978). *Stress.* London: Macmillan.

Cummins, R. A. (2003). Normative life satisfaction: Measurement issues and a homeostatic model. *Social Indicators Research*, 64(2), 225–256.

Cummins, R. A. (2005). Moving from the quality of life concept to a theory. *Journal of Intellectual Disability Research*, 49(10), 699–706.

Cummins, R. A. (2010). Subjective wellbeing, homeostatically protected mood and depression: A synthesis. *Journal of Happiness Studies*, 11(1), 1–17.

Cummins, R. A., Gullone, E. & Lau, A. L. (2002). A model of subjective well-being homeostasis: The role of personality. *In:* Gullone, E. & Cummins, R. A. (eds) *The Universality of Subjective Wellbeing Indicators.* Dordrecht: Springer, 7–46.

Curry Report (2002). *Farming and Food: A sustainable future.* Report of the Policy Commission on the Future of Farming and Food. London: Cabinet Office.

Day, G. (1998). A community of communities? Similarity and difference in Welsh rural community studies. *Economic and Social Review*, 29(3), 233–257.

Deary, I. J., Willock, J. & McGregor, M. (1997). Stress in farming. *Stress and Health*, 13(2), 131–136.

Defra (2007). *Sustainable Farming and Food Strategy – Indicator data sheet. 7.04: Farmer suicide rates.* [Online] Available at: http://webarchive.nationalarchives.gov.uk/20091105215609/https://statistics.defra.gov.uk/esg/indicators/d704_data.htm [accessed 02/10/17].

Defra (2010). *Public Attitudes to Agriculture, the Farmed Landscape and Natural Environment.* Discussion Paper prepared for the Agricultural Change and Environment Observatory. London: Defra.

Defra (2017). *Balance sheet analysis and farming performance, England 2015/2016 – dataset.* [Online] Available at: www.gov.uk/government/statistics/balance-sheet-analysis-and-farming-performance-england [accessed 17/10/17].

Defra (2018). *Rural Business Statistics: Rural businesses.* London: Defra.

Eberhardt, B. J. & Pooyan, A. (1990). Development of the farm stress survey: Factorial structure, reliability and validity. *Educational and Psychological Measurement*, 50(2), 393–402.

Firth, H. M., Williams, S. M., Herbison, G. P. & McGee, R. O. (2007). Stress in New Zealand farmers. *Stress and Health*, 23(1), 51–58.

Gasson, R. & Errington, A. J. (1993). *The Farm Family Business.* Wallingford: Cab International.

Gray, J. N. (2000). *At Home in the Hills: Sense of place in the Scottish Borders.* Oxford: Berghahn.

Hawton, K., Simkin, S., Malmberg, A., Fagg, J. & Hariss, L. (1998). *Suicide and Stress in Farmers.* London: The Stationery Office.

Health and Safety Executive (2016). *Health and Safety in Agriculture, Forestry and Fishing in Great Britain, 2014/15.* London: Health and Safety Executive.

Huppert, F. A., Marks, N., Clark, A., Siegrist, J., Stutzer, A., Vittersø, J. & Wahrendorf, M. (2009). Measuring well-being across Europe: Description of the ESS Well-being module and preliminary findings. *Social Indicators Research*, 91(3), 301–315.

Ilbery, B. W. (1998). *The Geography of Rural Change.* London: Longman.

Jones, P., Hawton, K., Malmberg, A. & Jones, J. W. (1994). Setting the scene. The background to stress in the rural community: Causes, effects and vulnerable groups. *In:* Read, N. (ed.) *Rural Stress: Positive action in partnership.* Stoneleigh Park: National Agriculture Centre.

Lobley, M., Johnson, G., Reed, M., Winter, M. & Little, J. (2004). *Rural Stress Review: Final Report*. Centre for Rural Policy Resesarch, University of Exeter.

Lobley, M., Potter, C., Butler, A., Whitehead, I. & Millard, N. (2005). *The Wider Social Impacts of Changes in the Structure of Agricultural Businesses*. University of Exeter: Centre for Rural Policy Research.

Mahon, M. (2007). New populations; shifting expectations: The changing experience of Irish rural space and place. *Journal of Rural Studies*, 23(3), 345–356.

Marsden, T. (1999). Rural futures: The consumption countryside and its regulation. *Sociologia Ruralis*, 39(4), 501–526.

Mazzola, J. J., Schonfeld, I. S. & Spector, P. E. (2011). What qualitative research has taught us about occupational stress. *Stress and Health*, 27(2), 93–110.

McGregor, M. J., Willcock, J. & Deary, I. J. (1995). Farm stress. *Farm Management*, 9, 57–65.

McGregor, M. J., Willock, J., Dent, B., Deary, I. J., Sutherland, A., Gibson, G., Morgan, O. & Grieve, B. (1996). Links between psychological factors and farmer decision making. *Farm Management*, 9(5), 228–239.

Milbourne, P. (2003). Hunting ruralities: Nature, society and culture in 'hunt countries' of England and Wales. *Journal of Rural Studies*, 19(2), 157–171.

Mind (2013). *Mind assesses research linking work with stress* [Online]. Available at: www.mind.org.uk/news-campaigns/news/work-is-biggest-cause-of-stress-in-peoples-lives/#.WdIVdVtSyUk [accessed 15/10/17].

Morris, C., Jarratt, S., Lobley, M. & Wheeler, R. (2016). *Baseline Farm Survey – Final Report*. Report for Defra project LM0302 Sustainable Intensification Research Platform Project 2: Opportunities and Risks for Farming and the Environment at Landscape Scales.

Murdoch, J. & Marsden, T. (1994). *Reconstituting Rurality: Class, community and power in the development process.* London: UCL Press.

Mzoughi, N. (2014). Do organic farmers feel happier than conventional ones? An exploratory analysis. *Ecological Economics*, 103, 38–43.

Office for National Statistics (2017a). *Annual Population Survey: Personal well-being in the UK: 2015 to 2016*. [Online] Available at: www.ons.gov.uk/peoplepopulationandcommunity/wellbeing/datasets/headlineestimatesofpersonalwellbeing.

Office for National Statistics (2017b). *Suicide by occupation, England: 2011 to 2015 – dataset*. [Online] Available at: www.ons.gov.uk/peoplepopulationandcommunity/birthsdeathsandmarriages/deaths/datasets/suicidebyoccupationenglandmaindatatables [accessed 15/07/17].

Olff, M., Koeter, M. W. J., Van Haaften, E. H., Kersten, P. H. & Gersons, B. P. R. (2005). Impact of a foot and mouth disease crisis on post-traumatic stress symptoms in farmers. *The British Journal of Psychiatry*, 186(2), 165–166.

Parry, J., Lindsey, R. & Taylor, R. (2005). *Farmers, Farm Workers and Work-related Stress: A report for the Health and Safety Executive*. London: Policy Studies Institute.

Peak District Rural Deprivation Forum (2004). *Hard Times: A research report into hill farming and farming families in the Peak District*. Hope Valley: PDRDF.

Phelps, C. (2001). *Stress in Farming in North Yorkshire*. WPS/01/06. Phase 1 Report. Sheffield: Health and Safety Laboratory.

Pugh, J. (1996). Crisis on the family farm: A matter of life and death. *In:* Carruthers, S. P. & Miller, F. A. (eds) *Crisis on the Family Farm: Ethics or economics?* Reading: Centre for Agricultural Strategy.

Raine, G. (1999). Causes and effects of stress on farmers: A qualitative study. *Health Education Journal*, 58(3), 259–270.

Reed, M., Lobley, M., Winter, M. & Chandler, J. (2002). *Family Farmers on the Edge: Adaptability and change in farm households*. University of Plymouth: Department of Land Use and Rural Management.

Roberts, F. (1950). Stress and the general adaptation syndrome. *British Medical Journal*, 2(4670), 104–105.

Samaritans (2003). *Stressed Out Survey*. London: Mori Omnibus.

Satsangi, M., Gallent, N. & Bevan, M. (2010). *The Rural Housing Question: Community and planning in Britain's countrysides.* Cambridge: Policy Press.

Scott, A., Christie, M. & Midmore, P. (2004). Impact of the 2001 foot and mouth disease outbreak in Britain: Implications for rural studies. *Journal of Rural Studies*, 20(1), 1–14.

Setten, G. (2005). Farming the heritage: On the production and construction of a personal and practised landscape heritage. *International Journal of Heritage Studies*, 11(1), 67–79.

Stark, C., Gibbs, D., Hopkins, P., Belbin, A., Hay, A. & Selvaraj, S. (2006). Suicide in farmers in Scotland. *Rural and Remote Health*, 6(1), 509.

Sutherland, V. J. & Cooper, C. L. (1990). *Understanding Stress: A psychological perspective for health professionals*. London: Chapman Hall.

The Prince's Countryside Fund (2017). *Who'd be a Farmer Today?* London: The Prince's Countryside Fund.

University of Cambridge (2012). *Farming loved but misunderstood, survey shows.* [Online]. Available at: www.cam.ac.uk/research/news/farming-loved-but-misunderstood-survey-shows/ [accessed 04/10/17].

Winter, M. (2003). Embeddedness, the new food economy and defensive localism. *Journal of Rural Studies*, 19(1), 23–32.

Winter, M. & Lobley, M. (2016). *Is there a Future for the Small Family Farm in the UK?* Report to the Prince's Countryside Fund. London: Prince's Countryside Fund.

Woods, M. (2003a). Conflicting environmental visions of the rural: Windfarm development in mid Wales. *Sociologia Ruralis*, 43.

Woods, M. (2003b). Deconstructing rural protest: The emergence of a new social movement. *Journal of Rural Studies*, 19(3), 309–325.

Woods, M. (2011). *Rural.* London: Routledge.

8 Farmers and the environment

Introduction

Few agriculturally related issues have been more challenging and led to so much agonising by policymakers, academics and farmers over the last two to three decades than environmental issues. Nearly 30 years ago the renowned environmental historian Donald Worster (1990), in seeking to establish environmental history as a central perspective within history, looking particularly through the lens of interactions between people and the environment, set out three levels of analytical inquiry. They need some adaptation for a focus on agriculture in the context of a contemporary social scientific analysis. But we think the framework works well especially as it prompts a broad approach in a field which has too often been the preserve of narrow specialisms whether of ecology in field studies of environmental change, or a range of particular approaches to understanding farmer behaviour through the variable lenses of economics, psychology, cultural geography, sociology and so forth.

First, Worster suggested a need to *understand nature* itself, 'consequently, the environmental historian must turn for help to a wide array of the natural sciences and must rely on their methodologies, sources, and evidence' (Worster, 1990, p.1090). In this chapter we seek to understand nature in the sense that we articulate the positives, what the economists call the positive *externalities*, of agriculture. This is where landscape features and habitats specific to particular farming systems have high 'natural' environmental value in human culture. There are many examples of this. Hedges, for example, if well managed, are rich in biodiversity. Originating as a practical method of land enclosure and livestock barrier, they are entirely a by-product of farming (Barnes and Williamson, 2006; Burel and Baudry, 1995; Pollard et al., 1974) yet in some sense mimic certain features of a pre-peopled environment. A layered, laid or steeped hedge (BTCV, 1975; de Geus and van Slobbe, 2003) is structurally akin to thickets of brambles and thorn within woodland undergrowth or areas of woodland cleared temporarily by windfall, and provides well protected nesting sites for birds and cover for small mammals such as hedgehogs. The abundance of other species is less a result of mimicry or relic but more of new opportunities. This is true for many farmland bird species such as the skylark which are dependent on large tracts of open, usually *cultivated*, land.

Second, Worster urged a focus on

> productive technology as it interacts with the environment... understanding how technology has restructured human ecological relations, that is, with analysing the various ways people have tried to make nature over into a system that produces resources for their consumption.
>
> (Worster, 1990, p.1090)

We move now to negative externalities as we examine the downside of agriculture, particularly as it intensifies and impacts negatively on environmental quality.

After sections in this chapter examining these two sides of the agriculture and environment coin, we turn in a third section to an examination of some of our own data on how farmers have responded to the environmental challenge and how agri-environmental land management is being incorporated in to everyday notions of what it means to be a 'good farmer'.

Worster's third analytic approach was a cultural level which examines 'perceptions, ideologies, ethics, laws, and myths (that) have become part of an individual's or group's dialogue with nature.' This seems to us to offer a helpful framing in the context of our primary concern for the social dimension of sustainable agriculture, and in this section we examine what and how farmers think about and relate to the environment.

Understanding nature: benign agriculture?

As indicated above, we are interested here in Britain's deep experience of farmed and managed landscapes. As Winter (2013) puts it,

> In long inhabited parts of the world, such as Europe, it is hard to conceive of rural environments that are not agricultural (or silvicultural). The very language we use testifies to the empirical and conceptual interaction of farming and the environment. We design *agri-environment* schemes, we talk of *agro-ecology*, we celebrate *farmland* birds and hedges. We place trust in organisations with such names as LEAF (Linking Environment and Farming) and FWAG (Farming and Wildlife Advisory Group) in Britain (Cox et al., 1990).
>
> (Winter, 2013, p.196; emphasis in original)

Thus the identification of High Nature Value (HNV) farming systems in Europe is evident in work emerging in the 1990s (Baldock et al., 1994; Bignal and McCracken, 1996) and in EU policy and attempts to develop suitable indicators thereafter (Strohbach et al., 2015). There has been a sustained attempt by a number of environmental organisations, such as the Institute of European Environmental Policy and the European Forum for Nature Conservation and Pastoralism, to develop the concept and to give it a stronger policy profile, resulting in the concept quickly becoming recognised under EU policy as a priority for rural development and halting biodiversity loss by 2020 (Strohbach et al., 2015). The monitoring of HNV farmland has been a requirement under the Common Monitoring Evaluation Framework of the EU Rural Development Programme, but national approaches to defining HNV systems have varied (Aue et al., 2014). Thus, for example, the Scottish government's (2011) approach is based on three characteristics of HNV. These are (i) the *presence of semi-natural vegetation*, such as unimproved grazing land, traditional hay meadows, semi-natural farmland features such as mature trees and uncultivated patches, or linear habitats such as streams and hedges; (ii) the *diversity of land cover*, in particular the need for a mosaic of land cover and use, such as cropland, fallow land, semi-natural vegetation and farmland features; and (iii) *low-intensity farming characteristics*. Like most national approaches, the Scottish indicators are based on defining HNV farming at the landscape scale, but attempts have also been made elsewhere to measure nature-values at the farm level (e.g. Boyle et al., 2015).

A problem with much of the agro-ecological research hitherto is that it cannot easily disentangle specific farming system effects, as opposed to rather more generic farming practices. To do that it is necessary to establish more precise relationships and causal mechanisms within joint production. This work is now beginning to happen. For example, research has shown how the creation of field margins can enhance the abundance and species richness of wildlife while maintaining, or even increasing, crop yield despite the loss of cropland for habitat creation (Pywell et al., 2015). These yield benefits can be explained by the 'spill-over of beneficial agro-ecological processes from adjacent wildlife habitats' (Pywell et al., 2015, p.5), such as increases in the number of crop pollinators and predatory insects that feed on pests. These and other research findings are continually being developed in order to investigate how ecological value through joint production can be maximised on farms. Thus Oppermann (2003) has sought to develop a 'nature balance scheme' for farms, and Smeding and de Snoo (2003) develop the concept of a food-web structure to research the ecological and biodiversity benefits and linkages of organic arable systems. Oppermann's 'nature balance scheme' provides farmers with a method of evaluating the ecological situation on their farm by recording data against criteria under four sectors (biodiversity landscape structure; biodiversity species richness; farm management and field management). Weaknesses in one sector can be compensated for by strengths in another, with the aim of the farmer achieving a 'balance' of 100 points across the board. Joint production is promoted in this case through 'enabl[ing] the assessment of the ecological situation for the farmers themselves, thus raising consciousness for ecological values and for ecologically sound practices' (Oppermann 2003 p.464). Focusing on the joint production capacity of organic farming, Smeding and de Snoo (2003) draw and compare food webs for five organic farms to examine the interlinkages between farming practices, ecosystem services and nature conservation. The authors conclude that organic farming can be useful for achieving the aims of both ecosystem services (of benefit to agriculture) and nature conservation, but that ecological food webs can vary significantly according to the intensity and duration of the organic farming system. Food-web complexities and dependencies also means that 'an optimisation of the farm food web with regard to ecosystem services, may possibly run counter to nature conservation goals' (Smeding and de Snoo, 2003, p.219). For instance, the increase of species beneficial for pest-control (i.e. polyphageous epigeic predators) may not support conservation-worthy species of insectivorous birds.

Technology and negative agriculture

Agriculture of course is not always or everywhere benign and from the 1960s, it became increasingly clear that agriculture's environmental consequences required much greater attention both in Europe and the 'neo-Europes' (Crosby, 1986) – Australia and New Zealand, North America and large parts of Latin America. It was the publication in 1962 of Rachel Carson's *Silent Spring* that first grabbed the headlines. Her exposure of the impact of agricultural pesticides and their global reach in food chains galvanised 'an embryonic environmental protest movement in the US and elsewhere in the world, although the origins of the concern over pesticides in the US can be traced back even earlier (Gunter and Harris, 1998)' (Winter, 2013, p.197). What originated as generalised concern over pesticides often translated into specific concerns in particular places. In Britain this crystallised around iconic species such as otters and birds of prey, as well as gamebirds (Sheail, 1985). At this stage, farmers were not necessarily being blamed for causing environmental problems. They may have been the agents of damage but as pesticides were the product of science and of industry, 'farming could be characterised as a victim as much as a perpetrator' (Winter, 2013, p.197). Consequentially, political campaigning was largely directed at improving the science behind pesticides (i.e. a shift away from broad spectrum products) and the stricter regulation and licensing of agro-chemical products. The tone for this was set by the decision by the Kennedy presidency to instruct the President's Science Advisory Committee, rather than the US Department of Agriculture, to investigate the issue and propose solutions (Wang, 1997).

However, over time, initial concern with the impact of pesticides was joined by a growing and powerful critique of the agricultural industry as evidence emerged of its ability to transform the countryside through a combination of the loss of landscape features and small-scale habitats; clearance and conversion of large blocks of semi-natural habitats; and a decline in traditional management and farming practices (Potter and Lobley, 1996). As evidence grew of the loss of ponds and hedgerows, for example, and the extent of the transformation of agricultural landscapes through 'improvement' and conversion of cherished habitats such as moorlands and chalk downland, early explanations that it was the actions of only a few 'maverick' farmers became hard to maintain and attention soon focused on the actions of farmers more generally, albeit aided and abetted by policy. The so-called policy thesis (Lowe et al., 1992) stated that the operation of the Common Agricultural Policy created conditions of confidence (through guaranteed prices – at the time – and investment aid) in which farmers were encouraged to specialise, increase output and intensify production through new technology embodied in capital, which was increasingly substituted for labour. Indeed, principles of investment and intensification, not to mention policies of agricultural support, were by now deeply engrained in the farming industry and consequently unparalleled changes were taking place. Farmers, once characterised as 'the nation's landscape gardeners' (Scott Committee, 1942, p.47), whose innate stewardship it was assumed would care for and protect the countryside, came under sustained attack from a number of British commentators (e.g. Body, 1982; Bowers and Cheshire, 1983; Lowe et al., 1986; Shoard, 1980). The degree of blame ascribed to farmers varied; while Marion Shoard famously castigated farmers for their 'theft of the countryside', Cheshire argued that

> the problem is not one of ill-will and ignorance but a system which systematically establishes financial inducements to erode the countryside, offers no rewards to prevent market failure and increases the penalties imposed ... on farmers who may want to farm in a way which enhances and enriches the rural environment.
>
> (Cheshire 1985, p.17)

Although the precise mechanisms and relative strength of different factors influencing this process of countryside change were debated, evidence of the pattern of change and associated environmental impacts was clear. This ranged from farm amalgamation – within England and Wales for example the average size of holding began sharply increasing from the mid-1960s, having been static for a century (Hine and Houston, 1973) – to enterprise specialisation at the expense of traditional mixed farming (Bowler, 1979). The extent of the loss of hedgerows and the ploughing up of flower rich meadows, all at tax-payers' expense as the cost of the Common Agricultural Policy escalated, became clear. For example, by the mid-1980s the area of species-rich, unimproved grassland was estimated to have declined by 97 per cent over the preceding 50 years (Fuller, 1987), while the six years from 1984–90 were estimated to have seen a 23 per cent net loss of hedgerows (Barr et al., 1993; 1994). The critique crystallised around a number of celebrated cases including the conversion of wetlands to arable agriculture in Halvergate and the Somerset Levels and the ploughing up for improved pasture of moorland on Exmoor (Lowe et al., 1986). Later, the organic movement built on some of this discontent by broadening the terms of the debate to include food quality and food safety alongside more traditional environmental and landscape concerns (Lockeretz, 2007).

Similar changes have been recorded throughout Europe. In Germany, for example, early evidence suggested that agriculture had directly or indirectly influenced the decline of approximately 87 per cent of all species (Conrad, 1991). In the Netherlands 55 per cent of wetlands were lost between 1950–85 (Baldock, 1990), while in Sweden a 66 per cent reduction in the number of farms between 1950–90 resulted in production becoming more concentrated and specialised (Vail, 1991) and is associated with a 50 per cent reduction in the area of meadow between 1950–80 (Pettersen, 1993) and the removal of 'hindrances to cultivation' (Kumm, 1991) – typically important landscape features and habitats. In addition to the loss of species, habitats and landscape features, growing recognition of the extent to which waterways were becoming contaminated by dissolved organic carbon, nitrate, phosphate and sediment run-off (Haygarth and Jarvis, 2002; Kay et al., 2009) means that the characterisation of the 'environmental problem' in agriculture has shifted during the past 20 years and the range of concerns is now very much greater, encompassing for instance biodiversity, food quality and safety, and water quality. For example, Pretty et al.'s (2000) estimates of the cost of treating drinking water to remove or reduce pesticides, nitrates, phosphorus and organic carbon (and related sediment) come to a total for the UK of £297 million.

Within this broad range of concerns the effect of agriculture on farmland bird populations is one of the more popularised issues regarding the negative externalities of agriculture, with sustained concern over declines in bird populations both in England and across Europe over the last 40 years. Thus we now have a history of research on the agricultural conditions associated with a number of different species (Donald, 1997; Hatchwell et al., 1996; Peach et al., 2004; Potts, 1997; Wilson et al., 1997). In England, while the decline in farmland bird populations in England has slowed since the 1990s, it is still on a downward trend overall; see Figure 8.1 (Defra, 2017). Specialist farmland species declined by 73 per cent between 1970 and 2016 and generalist species by 4 per cent, though some species have fared better than others. For instance, the grey partridge, turtle dove and tree sparrow have declined by 90 per cent or more, but the stock dove and goldfinch have more than doubled (Defra, 2017); see Figure 8.2.

Agricultural intensification is widely believed to be at least partly responsible for these farmland bird declines and there is evidence to support this. For instance, a recent study in Germany found that trends in farmland bird populations were negatively related to increasing maize cultivation and decreased crop diversity at the landscape scale (Jerrentrup et al., 2017). However, determining precise relations between agricultural conditions and the status of bird populations is fraught with difficulties, as is well demonstrated by Chamberlain et al. (2000) in work examining time series data (1962–95) for bird populations against a series of 31 separate agricultural variables representing crop areas, livestock numbers, fertiliser application, grass production and pesticide use. Analysis of the agricultural variables revealed, for example:

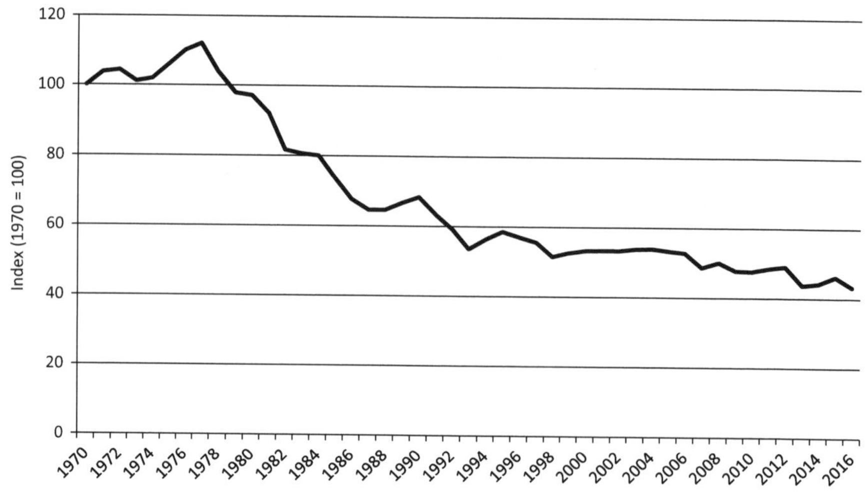

Figure 8.1 Farmland bird populations in England, 1970–2016

Source: Defra, 2017b

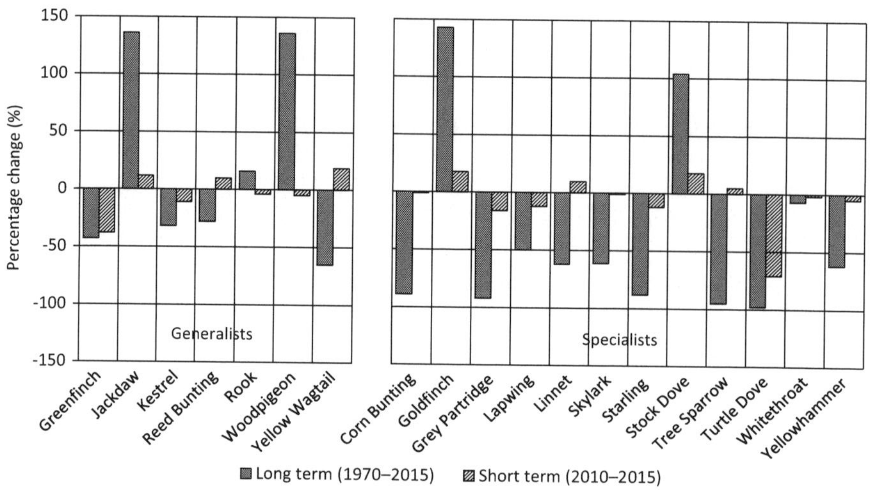

Figure 8.2 Percentage change in farmland bird species, long term (1970–2015) and short term (2010–2015)

Source: Defra, 2017b

- Large increases in the area of tilled land (by almost 1 million hectares between the early 1960s and 1995);
- Substantial changes in crop types and rotations (e.g. the replacement of root crops with oilseed rape, which has been enabled by the development of pre-emergent herbicides);
- A decrease in winter stubble due to a shift from spring to autumn-sown cereals;
- Declines in the area of all types of grasslands;
- The conversion of traditional hay meadows to grass grown for silage;
- Increases in the area of land treated with pesticide.

These changes represent a significant agricultural intensification from 1971 and coincide with a significant decline in the abundance of farmland birds, though the effects appear to have been lagged, with bird populations not declining until 1976.

While Chamberlain et al. (2000) are cautious not to claim precise causal links between specific agricultural practices and farmland bird decline, they do highlight some of the linkages between key agricultural variables and bird needs. For example, the use of pre-emergent and grass herbicides to facilitate autumn cereal sowing, along with declines in rough grazing and hay production, reduces the availability of food for some species of bird. Although now almost 20 years old, Chamberlain et al.'s (2000) work represents an important indicator of the growing realisation that subtle changes in agricultural practice, such the timing of cultivations or harvest, can have just as big an effect on wildlife as more visible changes, such as the removal of hedgerows, which had formed the focus of earlier campaigners.

Building on the benign and responding to the negative: farmers' environmental actions

Recognition of the environmental problems associated with agriculture, the CAP and the apparent erosion of the role of farmers as stewards of the countryside led to farmers facing increasing pressure to respond to the challenge. Numerous policies and campaigns (both government and third sector led) have been aimed at encouraging and supporting farmers to practice more 'environmentally friendly' farming, and consumer concern has also driven market-led initiatives such as farm assurance schemes, which usually have an environmental element, if not focus, to them.

Accordingly, the environmental activities of farmers has been the subject of extensive research (for overviews see Lastra-Bravo et al., 2015; Riley, 2011), with much of this work focusing on the effectiveness of agri-environmental policy and the factors affecting farmers' participation in formal agri-environment schemes (AES) (e.g. Burton, 2014; Lobley and Potter, 1998; Morris and Potter, 1995; Siebert et al., 2006; Wilson, 1996; Wilson and Hart, 2000; 2001). The development of AES is usually traced to the UK and a series of land use conflicts referred to above in the Norfolk broads, Somerset levels and elsewhere (see Baldock and Lowe, 1996; Potter, 1998). Although at the time a new idea, the notion that farmers could enter into contracts to provide public environmental goods quickly caught on. An 'agri-environmental consensus' between farming organisations and their former foes in the conservation movement began to emerge as the environmental importance of providing financial support for certain farming practices and farming systems became recognised. Beginning in the mid-1980s, AES have greatly helped the process of rehabilitating the notion of farmers as stewards of the countryside.

A large body of AES participation studies have highlighted the importance of financial factors and the details of scheme design in motivating AES participation (Lobley and Potter, 1998; Wilson and Hart, 2000). But research has also highlighted the complex interplay of social, cultural and psychological factors – for example, knowledge cultures (Morris, 2006), environmental values (Best, 2010) and self-identity constructs (Burton et al., 2008; Burton and Wilson, 2006; Lokhorst et al., 2014) – in shaping agri-environmental attitudes and behaviour. There is thus also a growing body of research that aims to understand farmer environmental behaviour with regard to voluntary, unsubsidised action, which has been argued to be more sustainable in the long-term than time-limited financially incentivised schemes (Barnes et al., 2013; van Dijk et al., 2016).

Here, we present some broad-level findings from the 2015 SIP Baseline Survey and 2006, 2010 and 2016 SW Farm Surveys around the environmental actions being taken by farmers. We focus particularly on 'formal' approaches to environmental land management (i.e. AES participation and documented management plans), while also touching on farmers' thoughts regarding the efficacy of such management tools. We do so to provide an illustrative snapshot of current actions being taken within the farming community, and to highlight some of the issues and sentiments expressed by farmers in relation to these.

Participation in AES

Evidence from the SIP Baseline Survey indicates that actions to address environmental issues are increasingly becoming a widespread and inescapable part of farming; whether driven by consumer demand, legislation, agri-environment payments, or personal concern for the environment on the part of the farmer. Strikingly, 78 per cent of the farmers in the 2015 SIP Baseline Survey held a current AES agreement (66 per cent of whom had also held a previous agreement), with related activities covering on average 64 per cent of the farm area. This high uptake is broadly comparable to national figures (for instance, in 2014, 70.7 per cent of utilisable agricultural area was in an AES (Natural England, 2014)) and perhaps reflects the importance of associated payments to many farmers, as discussed in Chapter 3. As Figure 8.3 shows, participation in AES in their various incarnations rose significantly throughout the 1990s and early 2000s, though has since declined,

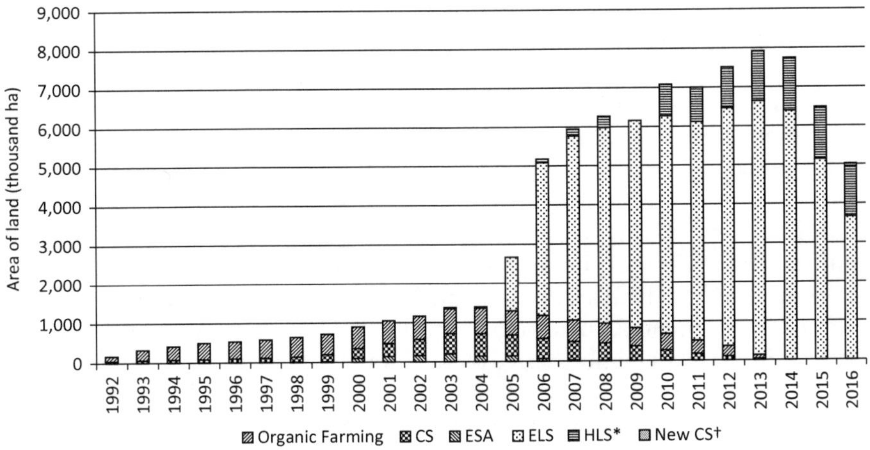

Figure 8.3 Area of land under agri-environment schemes in England, 1992–2016

Notes: CS – Countryside Stewardship.
ESA – Environmentally Sensitive Areas.
ELS – Environmental Stewardship Entry Level Scheme.
HLS – Environmental Stewardship Higher Level Scheme. *Includes Freestanding HLS and HLS linked to ELS.
New CS – New Countryside Stewardship. †Scheme opened in 2015 with first agreements going live in 2016. Includes mid and higher tier strands.

Source: Defra et al., 2017

with interest in the new Countryside Stewardship (CS) scheme (which opened for applications in 2015) reported as particularly low (NFU, 2015).

Although the majority of farmers in the SIP Baseline Survey held a current AES agreement, it is notable that, when asked about the main activities they undertook as part of this, the less demanding options (e.g. hedgerow management, low/no input areas and buffer strips) were particularly popular, in contrast to options requiring greater active management and/or change from pre-existing management practice (as has been noted elsewhere; see Lobley et al., 2013; Morris et al., 2000; Riley, 2006; Wilson and Hart, 2000). While this finding is unsurprising and not necessarily problematic in itself, it is potentially troubling when viewed alongside evidence from elsewhere which suggests that extrinsically motivated environmental action (i.e. action based on financial or legislative drivers), such as that fostered by AES, is not as effective at producing long-term, sustained benefits as voluntary action motivated by an intrinsic care for the environment (Barnes et al., 2013; Mills et al., 2017; van Dijk et al., 2016). Nevertheless, the government claims that agri-environment – or Environmental Land Management (ELM) – schemes have been 'beneficial to habitats and species, landscape character and water quality, with at least £3.20 of public goods returned for every £1 put in' (Defra and Government Statistical Service, 2018, p.59).

Certainly, the existence and design of AES has an impact on the way in which farmers farm and potentially on the wider changes they make to their business. In the 2006 SW Farm Survey, 39 per cent of farmers said their plans over the next five years (which might include changes to farmed area, livestock numbers, labour force or diversification activities, as well as environmental management) were influenced by environmental schemes, though this had dropped to 29 per cent in the 2016 survey; perhaps as a result of the introduction of the new CS scheme. Similarly, while 44 per cent of farmers in 2006, and 41 per cent in 2010, planned to increase their amount of environmental management (either formal or informal) over the next five years, only 22 per cent said the same in 2016 (1 per cent, 3 per cent and 5 per cent planned to decrease environmental management in 2006, 2010 and 2016 respectively, and the remainder planned no change or were not applicable).

There is a significant association with farm size here, with larger farms in the 2016 survey significantly more likely to have increased their environmental management since 2010 than smaller farms (53 per cent of farms over 250 hectares had done so compared to 19 per cent of those less than 50 hectares ($\chi^2=83.63$, $p<0.001$). Furthermore, in all three surveys larger farms were significantly more likely both to be planning increases in environmental management (e.g. in 2016, 27 per cent of farms over 250 hectares planned to increase it compared to 18 per cent of farms under 50ha; $\chi^2=39.78$, $p<0.001$) and to say their plans were influenced by environmental schemes (e.g. in 2016, 38 per cent of farms over 250 hectares were influenced by environmental schemes compared to 22 per cent of farms under 50 hectares; $\chi^2=16.48$, $p=0.006$[1]). As we saw in Chapter 6, the relationship between farm size and the environment is complex. The 'size effect' noted above may be a result of the greater opportunities to undertake formal and informal agri-environmental management on larger scale farms. Larger farms will also receive larger payments from agri-environmental schemes, which may explain why such schemes are more likely to influence the plans of larger farms.

Formal environmental management plans

Formal environmental management is, of course, not only limited to actions taken under AES but is also carried out and documented for other financial reasons (both subsidy and market related) and/or in accordance with legislation (e.g. cross-compliance and Nitrate Vulnerable Zone (NVZ) regulations). In addition to 78 per cent of farmers in the SIP Baseline Survey holding a current AES agreement, 26 per cent generated on-farm renewable energy (most commonly solar energy), 78 per cent were certified by at least one farm assurance scheme, and 93 per cent had at least one formal environmental management plan.[2] The extent to which these headline figures represent genuine concern and effective action regarding the environment is, however, questionable. While the majority of farmers reported having at least one environmental management plan, some admitted to not actively using them (see Table 8.1) and, of particular note, 48 per cent of those with a plan said that they had made no changes as a result. This is not to say that these farmers were not carrying out relevant environmental management activities, but rather that formal documentation – or the lack of it – should not necessarily be viewed as an indicator of whether or not effective environmental action is taking place. Several farmers appeared to view environmental plans as 'tick-box exercises' rather than as useful management tools, with some claiming either that they had been doing the activities anyway and/or that these plans do not accurately represent the extent and quality of relevant management. For example;

> You have a plan because you have to have one, it makes no difference to the farming behaviour. You're either aware of the issues and do something about them or you're not. Whether you have a plan or not is irrelevant.
>
> (Small lowland grazing livestock farmer, Taw)

> When you've been farming for as long as I have you just do things naturally, you don't need a plan. And also with livestock its sixth sense, you know if things aren't right when you go out there.
>
> (Medium lowland grazing livestock farmer, Upper Welland)

> There are so many of these things now that we have to do because we're required to, but I don't use them for farm management purposes really, because most of it we were doing anyway. Like the wildlife biodiversity management, it's a huge part of our business and we're doing it all the time, but it's not written on a piece of paper.
>
> (Very large cereal farmer, Avon)

Uptake of formal management plans also varied considerably by type, with 59 per cent of farmers actively using manure and nutrient plans but only 7 per cent using an energy efficiency plan (see Table 8.1). In the context of our discussion around the relative environmental performance of small farms in Chapter 6 (and further to the relationship between farm size and planned increases in environmental management noted above), it is also interesting that larger farms both tended to have more environmental plans (e.g. $\chi^2=34.89, p<0.001$ for manure management plans) and were more likely to have an AES agreement than small farms ($\chi^2=35.70, p<0.001$). Whether this is an indicator of greater environmental action, or simply of more extensive paperwork capabilities, is, however, difficult to ascertain.

Caring for the countryside

Regardless of the extent to which they were involved in formal environmental management, the farmers we interviewed generally professed to value the environment and certainly did not intentionally cause harm. This emerged in response to our open questions around farming's benefit and potential harm to the environment and local community. For instance, several participants stressed an ethos of care and *effort* (if not success) in this regard:

Table 8.1 Uptake of environmental management plans*

Type of plan	*Have and actively use* %	*Have but do not actively use* %	*Do not have a formal plan* %	*N/A to farming system* %
Manure	59	15	20	7
Nutrient	59	6	31	5
Energy efficiency	7	2	80	10
Crop protection/ integrated pest management	40	3	31	26
Soil	61	11	26	3
Wildlife/biodiversity	27	2	67	4
Water	18	3	64	15
Pollution risk assessment & abatement	45	7	39	9
Animal health	63	4	16	18
Other	8	0	1	1

Note: *Columns do not sum to 100% as respondents may have more than one management plan.

Source: Morris et al., 2016

> I like to see my swallows, my birds, all my other animals, so if I can manage around them and not cut that field until they've nested and gone, then I do that. So it's not my single aim, but that's all part of farming isn't it, looking after it as best you can for as long as you can.
>
> (Medium lowland grazing livestock farmer, Taw)

> All I can say is we try not to be detrimental. In farming you can't do more than try. I like to think we're environmentally aware, put it that way. I'm not saying we always get it right but we do try.
>
> (Very large cereal farmer, Upper Welland)

The theme of stewardship arose frequently in these discussions and many farmers commented on the importance of farming to the upkeep of the countryside. For instance:

> We're providing the backdrop to the landscape, to the environment, that invites tourists to enjoy the area, so we're maintaining the environment which makes Norfolk what it is. So that's part of it … It's maintaining the environment and producing food – which are sustainable and go hand-in-hand.
>
> (Ultra-large general cropping farmer, Wensum & Yare)

> If farmers didn't farm the land it would just be a barren wilderness.
>
> (Large mixed farmer, Avon)

In a similar vein, the 2016 SW Farm Survey found that 64.1 per cent of respondents strongly agreed and 20.6 per cent agreed with the statement 'Farming is important in maintaining the local environment'. Such identification of farmers as 'custodians of the countryside' has been shown to positively influence farmer attitudes towards the environment (Mills et al., 2013), but since the custodianship/stewardship ideology tends to emphasise landscape, or 'god's garden', ideals, its objectives do not necessarily align with ecological success (Fish et al., 2003). Ellis (2013) also points out that custodianship's 'symbiotic ideology', which asserts the ability for farmers to simultaneously achieve both production and environmental goals, acts as an 'ideological trick' which allows farmers to maintain a self-narrative of sustainability, regardless of actual environmental success (see also Kessler et al., 2016). There is thus a danger of inaction where farmers believe they are already 'environmentally friendly'. The fact that 68 per cent of those in the SIP Baseline Survey did not believe their farming activities had an *avoidable* detrimental impact on the environment (18 per cent admitted they did, 13 per cent were unsure) may be telling in this regard.

A lack of clarity among farmers regarding the extent of agriculture's impact on the environment has been observed elsewhere. For example, in relation to diffuse water pollution, Inman et al. (2018) note that farmers in their discussion groups were confused about the scale of the problem, and their ability to make a tangible difference to it, partly because they had not been presented with scientific evidence about the condition of their local waterbodies. Importantly, our findings from the SIP Baseline Survey suggest that where farmers do acknowledge environmental harm this is certainly not accompanied by a lack of concern. For instance, one farmer explained how

> Despite what we do in terms of margins and targeted input of applications, here we are on a sloping part of the valley, so we must be getting run-off into water courses … It does concern me.
>
> (Medium cereal farmer, Upper Welland)

Such admissions highlight the need to avoid painting farmers as uncaring and irresponsible – particularly since, as we saw in Chapter 7, farmers often feel undervalued and 'blamed' for environmental ills – but to instead further support and enable them to improve their environmental performance. As asserted elsewhere in the literature, the *ability* to adopt environmental measures is as crucial as the *willingness* to do so (Mills et al., 2017; Siebert et al., 2006) and this depends on a multiplicity of factors, including business resources (e.g. finance, time) (Dwyer et al., 2007), farmer knowledge (Lobley et al., 2013) and the biogeographical nature of the land (Wilson and Hart, 2001).

Agriculture as culture

We move now to Donald Worster's (1990), third analytic level:

> that more intangible, purely mental type of encounter in which perceptions, ideologies, ethics, laws, and myths have become part of an individual's or group's dialogue with nature. People are continually constructing cognitive maps of the world around them, defining what a resource is, determining which sorts of behaviour may be environmentally degrading and ought to be prohibited, and generally choosing the ends to which nature is put.
>
> (Worster, 1990, p.1091)

Given our concern with the 'social' dimension of food and farming, in this section we examine what and how farmers think about and relate to the environment before moving on to consider some of the ways in which (re)connections have been made between food, farmers, the environment and consumers.

Shifts in farming identities?

The findings we have presented in the previous section suggest that most farmers recognise the need to incorporate formal environmental management into their farming practices, at least to some extent, even if only to 'tick the boxes' for cross-compliance or farm assurance schemes, or access agri-environment payments. The extent to which AES and other policies and campaigns have led to genuine change in the 'conservation ethos' of the farming community is, however, more difficult to ascertain. A sustained concern within much of the literature on farmer cultures is that deep-seated 'productivist' identities are slow to change and that a traditional emphasis on 'tidy' and productive landscapes continues to conflict with ecological goals and inhibit conservation action (Burton, 2004; Burton et al., 2008; Burton and Wilson, 2006; Egoz et al., 2001; Fish et al., 2003; Kings and Ilbery, 2010; McEachern, 1992; Silvasti, 2003). On the other hand, there is some evidence that the 'conservationist' element of farmer attitudes, identities and actions has strengthened in response to the environmental challenge and that social norms are beginning to alter (at least in some contexts), partly due to external drivers such as changing economic and market conditions (Saunders, 2016; Sutherland and Darnhofer, 2012), AES (Huttunen and Peltomaa, 2016; Riley, 2016) and other agri-environment initiatives (McGuire et al., 2013; van Dijk et al., 2015). For example, in a case study of hill farmers in the Peak District, Riley (2016) shows how long-term participation in AES – and the assimilation of new scientific understandings into farmer knowledge cultures as part of this – led to AES activities being accommodated into regional concepts of 'good farming'.

Inman et al. (2018) point out that identities are more likely to change if the social environment is dynamic and suggest that access to new networks and ideas can help foster conservation-oriented identities. They advocate the use of 'double loop learning' through a group learning processes whereby actions are reviewed, evaluated and subsequently modified in order to develop new ways of thinking and, ultimately, new social norms within the group. If appropriately designed, collaborative approaches to environmental management – which have gained attention as potentially effective ways of achieving environmental improvements at the landscape scale (Carmona-Torres et al., 2011; Cong et al., 2014; Goldman et al., 2007; McKenzie et al., 2013) – may thus hold promise for initiating socially endorsed actions within farmer groups, though the complexities and challenges associated with such joint working are not insignificant (see Jarratt et al., 2015).

The changing nature of environmental concern in agriculture

Longstanding concern with agriculture's role in the fate of landscape, species and habitats has not gone away. Indeed, as we have seen, although at one point the notion of stewardship in agriculture may well have lain in tatters, farmers are now financially rewarded for their stewardship of the countryside through formal agri-environmental management contracts. In addition, there is widespread evidence of the 'informal' work undertaken by farmers in managing the environment (e.g. Mills et al., 2013). This shift of environmental concern to the heart of agricultural practice is also reflected in a growing concern for food quality defined by its environmental provenance (Morris and Kirwan, 2010). The debate is no longer just between farmers and conservationists. A new vocabulary has emerged to capture the relationship between the choices consumers make about food and environmental outcomes on the farm, hence terms like 'eating the view' and 'eating biodiversity'. This is part of a developing trend to harness the market to help deliver good environmental management and good quality food. Consumer concerns and demands have seen quality assurance schemes developed by retailers as part of the regulatory framework (Morris and Young, 2000) and branding and identity have been developed increasingly by farmers seeking to reconnect with consumers (Morris and Buller, 2003). As Winter puts it:

> [R]econnection is to do with making the links between particular foods and particular natures, a reterritorialization or respatialization of food production which begins to reverse the aspatialities which are, or were, an intrinsic part of a globalized food order. … It is also to do with a growing realisation that the properties of food are 'natural' properties and that heterogeneity of edaphic conditions gives rise to varied natures represented in varied foods.
>
> (Winter, 2005, p.611)

Given the interplay of the physical and social in the origins and production of foods and, indeed, implicit in the notion of agri-culture, this provides a rich territory for social scientific endeavour around hybridity and materialities (Anderson and Tolia-Kelly, 2004; Kirsch and Mitchell, 2004; Whatmore, 2002).

There are many strands to this 'cultural' work but the emphasis here is on those who have sought to consider the implications of reconnection for measuring and improving environmental performance. Buller and Morris (2004), for example, explore the role of markets in attempts 'to reconcile agricultural production and environmental protection as new forms of commodification permit a shift in the values attributed to the various "products" of agricultural enterprise' (p.1067). A particularly useful contribution is their attempt to relate market-oriented initiatives in the agro-food sector back to other key debates in agro-food studies in a manner not undertaken in earlier approaches to the 'quality turn'. Rather than subscribe to an 'either/or' approach to market and policy approaches, Buller and Morris side with Bourdieu in the assertion that 'the economic field is inhabited by the state which contributes, at each and every moment, to its existence and its durability' (Bourdieu, 2000, p.25). They also consider the way in which the marketing of products from sustainable food production systems may help to 'internalise' some of the negative externalities which, as we have seen, lie at the heart of the agri-environmental problem.

Although there are undoubtedly opportunities to harness the power of the market to deliver improved environmental performance in agriculture, Guthman (2004) is less sanguine in the context of agri-business penetration of the organic sector in California and the 'coventionalisation' that appears to be taking place. The so-called conventionalisation of organic farming and marketing describes the process by which, over time, organic farming systems, structures and supply chains increasingly take on characteristics of the conventional system to which organics is so often seen to be in opposition to. It is much debated both as a theoretical concept and an observable empirical trend (e.g. Best, 2008; Cakirli et al., 2017; Dantsis et al., 2009; Darnhofer et al., 2010; Goldberger, 2011;Hall and Mogyorody, 2001; Guptill, 2009; Lockie and Halpin, 2005; Ramos García et al., 2018; Schewe, 2014; Sutherland, 2013; Świergiel et al., 2018; Tovey, 2002). Evidence from this literature suggests that processes of conventionalisation are operating throughout the world, although several writers argue that the situation is both more nuanced and multi-directional than the concept implies (e.g. Best, 2008; Guptill, 2009; Lockie and Halpin, 2005; Schewe, 2014; Sutherland, 2013). Moreover, it has been argued that by establishing a dualism between the 'ideal' and 'good' alternative and the 'bad' conventional, conventionalisation fails to recognise positive developments towards more sustainable forms of agriculture and consumption, and fails to give credence to the diversity of 'organic' (Campbell and Rosin, 2011; Rosin and Campbell, 2009; Watts et al., 2018).

Conclusions

It is perhaps unsurprising in a chapter that has adapted a framework developed by an environmental historian that we have tracked much of the history of agriculture–environment relations over the last 40–50 years. From the conception of a benign agriculture underpinning our countryside; the powerful environmental critique of agriculture, with its origins in the 1960s, which culminated in the introduction of formal agri-environmental schemes in the mid-1980s; through to attempts to harness forces of the market to deliver an agriculture rich in environment and quality; food, farms and farmers remain at the centre of concern. This is not just the story of the rehabilitation of the stewardship role of farmers under a more conducive policy environment. As we have seen, the market is increasingly important in responding to the very broadly conceived relationship between farmers and the environment. This is not intended to suggest that we should uncritically engage with market forces nor that environment and agriculture conflicts are a thing of the past. That is clearly not true. In a small crowded group of islands like the UK, agriculture is always going to be implicated in debates over the countryside for, as we argued at the beginning of this chapter, in many ways agriculture is the countryside.

To us, a major implication to flow from this is the re-discovery of the importance of place (although some of us have never forgotten that place matters) and, equally important, the re-discovery of local study as the underpinning of an interdisciplinary understanding of agri-environmental issues. Recent years have witnessed a neglect of local sociologies of agriculture with remarkably few local studies of agrarian sociology following in the path of Howard Newby's seminal work of the 1970s. We argue that there is a need for studies that will give due weight to local social formations and local politics, to farmers' worldviews within that local context, to the interactions between farmer behaviour and their occupancy and use of land, to the interactions farmers have, through the food chain, with consumers and wider publics, and to the ecosystem services that flow from our use of the land. There is no shortage of social surveys of farmers but sustained scholarship within a specific geographical milieu is today very rare and needs to be revived as we seek to understand local natures, local production systems and local cultures in their global contexts.

Notes

1 This relationship was also present in 2006 ($\chi2=53.06$, $p<0.001$) and 2010 ($\chi2=32.83$, $p<0.001$).

2 Farmers were asked if they had formal plans concerning the management of: manure, nutrients, energy efficiency, crop protection, soil, wildlife, water, pollution risk, animal health or 'other'.

References

Anderson, B. & Tolia-Kelly, D. (2004). Matter(s) in social and cultural geography. *Geoforum*, 35(6), 669–674.

Aue, B., Diekötter, T., Gottschalk, T. K., Wolters, V. & Hotes, S. (2014). How High Nature Value (HNV) farmland is related to bird diversity in agro-ecosystems – Towards a versatile tool for biodiversity monitoring and conservation planning. *Agriculture, Ecosystems & Environment*, 194, 58–64.

Baldock, D. (1990). *Agriculture and Habitat Loss in Europe*. CAP Discussion Paper No.3. London: WWF International.

Baldock, D., Beaufoy, G. & Clark, J. (1994). *The Nature of Farming: Low intensity farming systems in nine European countries*. London: Joint Nature Conservation Committee.

Baldock, D. & Lowe, P. (1996). The development of European agri-environmental policy. *In:* Whitby, M. (ed.) *The European Environment and CAP Reform: Policies and prospects for conservation*. Wallingford: CAB International, 8–25.

Barnes, A. P., Toma, L., Willock, J. & Hall, C. (2013). Comparing a 'budge' to a 'nudge': Farmer responses to voluntary and compulsory compliance in a water quality management regime. *Journal of Rural Studies*, 32, 448–459.

Barnes, G. & Williamson, T. (2006). *Hedgerow History: Ecology, history and landscape character*. Macclesfield: Windgather Press.

Barr, C., Bunce, R. G. H., Clarke, R., Fuller, R., Furse, M., Gillespie, M., Groom, G., Hallman, C., Hornung, M., Howard, D. & Ness, M. (1993). *Countryside Survey 1990: Main report*. CS 1990 Volume 2. London: Department of the Environment

Barr, C., Gillespie, M. & Howard, D. (1994). *Hedgerow Survey 1993: Stock and change estimates of hedgerow lengths in England and Wales 1990–93*. Contract report to the DOE.

Best, H. (2008). Organic agriculture and the conventionalization hypothesis: A case study from West Germany. *Agriculture and Human Values*, 25(1), 95–106.

Best, H. (2010). Environmental concern and the adoption of organic agriculture. *Society & Natural Resources*, 23(5), 451–468.

Bignal, E. M. & McCracken, D. I. (1996). Low-intensity farming systems in the conservation of the countryside. *Journal of Applied Ecology*, 413–424.

Body, R. (1982). *Agriculture: The triumph and the shame*. London: Temple Smith.

Bourdieu, P. (2000). *Les Structures Sociales de L' Économie*. Paris: Éditions du Seuil.

Bowers, J. K. & Cheshire, P. (1983). *Agriculture, The Countryside and Land Use: An economic critique*. London: Methuen.

Bowler, I. (1979). *Government and Agriculture: A spatial perspective*. London: Longman.

Boyle, P., Hayes, M., Gormally, M., Sullivan, C. & Moran, J. (2015). Development of a nature value index for pastoral farmland – A rapid farm-level assessment. *Ecological Indicators*, 56, 31–40.

BTCV (1975) *Hedging*. London: British Trust for Conservation Volunteers.

Buller, H. & Morris, C. (2004). Growing goods: The market, the state, and sustainable food production. *Environment and Planning A: Economy and space*, 36(6), 1065–1084.

Burel, F. & Baudry, J. (1995). Social, aesthetic and ecological aspects of hedgerows in rural landscapes as a framework for greenways. *Landscape and Urban Planning*, 33(1–3), 327–340.

Burton, R. J. F. (2004). Seeing through the 'Good Farmer's' eyes: Towards developing an understanding of the social symbolic value of 'productivist' behaviour. *Sociologia Ruralis*, 44(2), 195–215.

Burton, R. J. F. (2014). The influence of farmer demographic characteristics on environmental behaviour: A review. *Journal of Environmental Management*, 135, 19–26.

Burton, R. J. F., Kuczera, C. & Schwarz, G. (2008). Exploring farmers' cultural resistance to voluntary agri-environmental schemes. *Sociologia Ruralis*, 48(1), 16–37.

Burton, R. J. F. & Wilson, G. (2006). Injecting social psychology theory into conceptualisations of agricultural agency: Towards a post-productivist farmer self-identity? *Journal of Rural Studies*, 22(1), 95–115.

Cakirli, N., Lakner, S., Theuvsen, L. & Otter, V. (2017) Adoption of organic agriculture: An exploration of the factors influencing conversion to organic farming in the light of the conventionalization debate. *Agrarian Perspectives XXVI. Competitiveness of European Agriculture and Food Sectors, Proceedings of the 26th International Conference, 13–15 September 2017 Prague, Czech Republic*. Czech University of Life Sciences Prague, Faculty of Economics and Management, 31–39.

Campbell, H. & Rosin, C. (2011). After the 'Organic Industrial Complex': An ontological expedition through commercial organic agriculture in New Zealand. *Journal of Rural Studies*, 27(4), 350–361.

Carmona-Torres, C., Parra-López, C., Groot, J. C. J. & Rossing, W. A. H. (2011). Collective action for multi-scale environmental management: Achieving landscape policy objectives through cooperation of local resource managers. *Landscape and Urban Planning*, 103(1), 24–33.

Chamberlain, D. E., Fuller, R. J., Bunce, R. G. H., Duckworth, J. C. & Shrubb, M. (2000). Changes in the abundance of farmland birds in relation to the timing of agricultural intensification in England and Wales. *Journal of Applied Ecology*, 37(5), 771–788.

Cheshire, P. (1985). The environmental implications of European agricultural support policies. *In:* Baldock, D. & Conder, D. (eds) *Can the CAP Fit the Environment?* London: CPRE, IEEP, 9–18.

Cong, R.-G., Smith, H. G., Olsson, O. & Brady, M. (2014). Managing ecosystem services for agriculture: Will landscape-scale management pay? *Ecological Economics*, 99, 53–62.

Conrad, J. (1991). *Options and Restrictions of Environmental Policy in Agriculture.* Baden-Baden: Nomos Verlag.

Cox, G., Lowe, P. & Winter, M. (1990). *The Voluntary Principle in Conservation: A study of the Farming and Wildlife Advisory Group.* Chichester: Packard.

Crosby, A. W. (1986). *Ecological Imperialism: The biological expansion of Europe, 900–1900.* Cambridge: Cambridge University Press.

Dantsis, T., Loumou, A. & Giourga, C. (2009). Organic agriculture's approach towards sustainability; its relationship with the agro-industrial complex, a case study in central Macedonia, Greece. *Journal of Agricultural and Environmental Ethics*, 22(3), 197–216.

Darnhofer, I., Lindenthal, T., Bartel-Kratochvil, R. & Zollitsch, W. (2010). Conventionalisation of organic farming practices: From structural criteria towards an assessment based on organic principles. A review. *Agronomy for Sustainable Development*, 30(1), 67–81.

Defra (2017). *ENV08 – Wild bird populations in England.* [Online] Available at: www.gov.uk/government/statistical-data-sets/env08-wild-bird-populations-in-england [accessed 30/04/18].

Defra, Department of Agriculture Environment and Rural Affairs (Northern Ireland), Welsh Assembly Government & The Scottish Government (2017). *Agriculture in the United Kingdom 2016*. London: Defra.

Defra & Government Statistical Service (2018). *The Future Farming and Environment Evidence Compendium*. London: Defra.

Donald, P. F. (1997). The corn bunting *Milaria calandra* in Britain: A review of current status, patterns of decline and possible causes. *In:* Donald, P. F. & Aebischer, N. J. (eds) *The Ecology and Conservation of Corn Buntings Milaria calandra.* Peterborough: Joint Nature Conservation Committee, 11–26.

Dwyer, J., Mills, J., Ingram, J., Taylor, J., Burton, R., Blackstock, K., Slee, B., Brown, K., Schwartz, G., Matthews, K. & Dilley, R. (2007). *Understanding and Influencing Positive Behaviour Change in Farmers and Land Managers – a project for Defra.* Gloucester: CCRI, Macaulay Institute.

Egoz, S., Bowring, J. & Perkins, H. C. (2001). Tastes in tension: Form, function, and meaning in New Zealand's farmed landscapes. *Landscape and Urban Planning*, 57, 177–196.

Ellis, C. (2013). The symbiotic ideology: Stewardship, husbandry, and dominion in beef production. *Rural Sociology*, 78(4), 429–449.

Fish, R., Seymour, S. & Watkins, C. (2003). Conserving English landscapes: Land managers and agri-environmental policy. *Environment and Planning A*, 35(1), 19–41.

Fuller, R. M. (1987). The changing extent and conservation interest of lowland grasslands in England and Wales: A review of grassland surveys 1930–1984. *Biological Conservation*, 40(4), 281–300.

de Geus, M. and van Slobbe, T. (2003) *Fences and Freedom: The philosophy of hedge laying*. Utrecht: International Books.

Goldberger, J. R. (2011). Conventionalization, civic engagement, and the sustainability of organic agriculture. *Journal of Rural Studies*, 27(3), 288–296.

Goldman, R. L., Thompson, B. H. & Daily, G. C. (2007). Institutional incentives for managing the landscape: Inducing cooperation for the production of ecosystem services. *Ecological Economics*, 64(2), 333–343.

Gunter, V. J. & Harris, C. K. (1998). Noisy winter: The DDT controversy in the years before Silent Spring. *Rural Sociology*, 63(2), 179–198.

Guptill, A. (2009). Exploring the conventionalization of organic dairy: Trends and counter-trends in upstate New York. *Agriculture and Human Values*, 26(1), 29–42.

Guthman, J. (2004). The trouble with 'organic lite' in California: A rejoinder to the 'conventionalisation' debate. *Sociologia Ruralis*, 44(3), 301–316.

Hall, A. & Mogyorody, V. (2001). Organic Farmers in Ontario: An examination of the conventionalization argument. *Sociologia Ruralis*, 41(4), 399–322.

Hatchwell, B. J., Chamberlain, D. E. & Perrins, C. M. (1996). The demography of blackbirds *Turdus merula* in rural habitats: Is farmland a sub-optimal habitat? *Journal of Applied Ecology*, 33, 1114–1124.

Haygarth, P. M. & Jarvis, S. (eds) (2002). *Agriculture, Hydrology and Water Quality*. Wallingford: CAB International.

Hine, R. C. & Houston, A. M. (1973). *Government and Structural Change in Agriculture*. Joint Report prepared by Universities of Nottingham and Exeter.

Huttunen, S. & Peltomaa, J. (2016). Agri-environmental policies and 'good farming' in cultivation practices at Finnish farms. *Journal of Rural Studies*, 44, 217–226.

Inman, A., Winter, M., Wheeler, R., Vain, E., Lovett, A., Collins, A., Jones, I., Johnes, P. & Cleasby, W. (2018). An exploration of individual, social and material factors influencing water pollution mitigation behaviours within the farming community. *Land Use Policy*, 70, 16–26.

Jarratt, S., Morris, C., Wheeler, R. & Winter, M. (2015). *Literature Review on Farming Collaboration.* Report for Defra project LM0302 Sustainable Intensification

Research Platform Project 2: Opportunities and Risks for Farming and the Environment at Landscape Scales.
Jerrentrup, J. S., Dauber, J., Strohbach, M. W., Mecke, S., Mitschke, A., Ludwig, J. & Klimek, S. (2017). Impact of recent changes in agricultural land use on farmland bird trends. *Agriculture, Ecosystems & Environment*, 239, 334–341.
Kay, P., Edwards, A. C. & Foulger, M. (2009). A review of the efficacy of contemporary agricultural stewardship measures for ameliorating water pollution problems of key concern to the UK water industry. *Agricultural Systems*, 99(2–3), 67–75.
Kessler, A., Parkins, J. R. & Huddart Kennedy, E. (2016). Environmental harm and "the good farmer": Conceptualizing discourses of environmental sustainability in the beef industry. *Rural Sociology*, 81(2), 172–193.
Kings, D. & Ilbery, B. (2010). The environmental belief systems of organic and conventional farmers: Evidence from central-southern England. *Journal of Rural Studies*, 26(4), 437–448.
Kirsch, S. & Mitchell, D. (2004). The nature of things: Dead labor, nonhuman actors, and the persistence of Marxism. *Antipode*, 36(4), 687–705.
Kumm, K. (1991). The effects of Swedish price support, fertiliser and pesticide policies on the environment. *In:* Young, M. (ed.) *Towards Sustainable Agricultural Development*. London: Belhaven, 50–90.
Lastra-Bravo, X. B., Hubbard, C., Garrod, G. & Tolón-Becerra, A. (2015). What drives farmers' participation in EU agri-environmental schemes?: Results from a qualitative meta-analysis. *Environmental Science & Policy*, 54, 1–9.
Lobley, M. & Potter, C. (1998). Environmental Stewardship in UK agriculture: A comparison of the environmentally sensitive area programme and the Countryside Stewardship Scheme in South East England. *Geoforum*, 29(4), 413–432.
Lobley, M., Saratsi, E., Winter, M. & Bullock, J. (2013). Training farmers in agri-environmental management: The case of Environmental Stewardship in lowland England. *International Journal of Agricultural Management*, 3(1), 12–20.
Lockeretz, W. (ed.) (2007). *Organic Farming: An international history*. Wallingford: CABI.
Lockie, S. & Halpin, D. (2005). The 'conventionalisation' thesis reconsidered: Structural and ideological transformation of Australian organic agriculture. *Sociologia Ruralis*, 45(4), 284–307.
Lokhorst, A. M., Hoon, C., le Rutte, R. & de Snoo, G. (2014). There is an I in nature: The crucial role of the self in nature conservation. *Land Use Policy*, 39(July), 121–126.
Lowe, P., Cox, G., MacEwen, M., O'Riordan, T. & Winter, M. (1986). *Countryside Conflicts. The politics of farming, forestry and conservation.* Aldershot: Gower and Maurice Temple Smith.
Lowe, P., Ward, N. & Munton, R. (1992). Social analysis of land use change: The role of the farmer. *In:* Whitby, M. (ed.) *Land Use Change: The causes and consequences.* London: ITE Symposium, HM Stationery Office, 42–51.
McEachern, C. (1992). Farmers and conservation: Conflict and accommodation in farming politics. *Journal of Rural Studies*, 8(2), 159–171.
McGuire, J., Morton, L. W. & Cast, A. D. (2013). Reconstructing the good farmer identity: Shifts in farmer identities and farm management practices to improve water quality. *Agriculture and Human Values*, 30(1), 57–69.
McKenzie, A. J., Emery, S., Franks, J. & Whittingham, M. (2013). Landscape-scale conservation: Collaborative agri-environment schemes could benefit both

biodiversity and ecosystem services, but will farmers be willing to participate? *Journal of Applied Ecology*, 50(5), 1274–1280.

Mills, J., Gaskell, P., Ingram, J., Dwyer, J., Reed, M. & Short, C. (2017). Engaging farmers in environmental management through a better understanding of behaviour. *Agriculture and Human Values*, 34, 283–299.

Mills, J., Gaskell, P., Reed, M., Short, C. J., Ingram, J., Boatman, N., Jones, N., Conyers, S., Carey, P., Winter, M. & Lobley, M. (2013). *Farmer attitudes and evaluation of outcomes to on-farm environmental management.* Report to Defra. Gloucester: CCRI.

Morris, C. (2006). Negotiating the boundary between state-led and farmer approaches to knowing nature: An analysis of UK agri-environment schemes. *Geoforum*, 37(1), 113–127.

Morris, C. & Buller, H. (2003). The local food sector: A preliminary assessment of its form and impact in Gloucestershire. *British Food Journal*, 105(8), 559–566.

Morris, C. & Kirwan, J. (2010). Food commodities, geographical knowledges and the reconnection of production and consumption: The case of naturally embedded food products. *Geoforum*, 41(1), 131–143.

Morris, C. & Potter, C. (1995). Recruiting the new conservationists: Farmers' adoption of agri-environmental schemes in the UK. *Journal of Rural Studies*, 11(1), 51–63.

Morris, C. & Young, C. (2000). 'Seed to shelf', 'teat to table', 'barley to beer' and 'womb to tomb': Discourses of food quality and quality assurance schemes in the UK. *Journal of Rural Studies*, 16(1), 103–115.

Morris, J., Mills, J. & Crawford, I. M. (2000). Promoting farmer uptake of agri-environment schemes: The Countryside Stewardship Arable Options Scheme. *Land Use Policy*, 17(3), 241–254.

Natural England (2014). *Land Management Update. June 2014*. Sheffield: Natural England

NFU (2015). *CS scheme 'too complex' – survey reveals* [Online]. Available at: www.nfuonline.com/news/latest-news/cs-scheme-too-complex-survey-reveals/ [accessed 11/10/17].

Oppermann, R. (2003). Nature balance scheme for farms—evaluation of the ecological situation. *Agriculture, Ecosystems & Environment*, 98(1–3), 463–475.

Peach, W., J., Denny, M., Cotton Pete, A., Hill Ian, F., Gruar, D., Barritt, D., Impey, A. & Mallord, J. (2004). Habitat selection by song thrushes in stable and declining farmland populations. *Journal of Applied Ecology*, 41(2), 275–293.

Pettersen, O. (1993). Scandinavian agriculture in a changing environment. *In:* Harper, S. (ed.) *The Greening of Rural Policy: International perspectives.* London: Belhaven, 82–98.

Pollard, E., Hooper, M. D. & Moore, N. W. (1974). *Hedges.* London: Collins.

Potter, C. (1998). *Against the Grain: Agri-environmental reform in the United States and the European Union.* Wallingford: CAB International.

Potter, C. & Lobley, M. (1996). The farm family life cycle, succession paths and environmental change in Britain's countryside. *Journal of Agricultural Economics*, 47(1–4), 172–190.

Potts, D. (1997). Cereal farming, pesticides and grey partridges. *In:* Pain, D. J. & Pienkowski, M. W. (eds) *Farming and Birds in Europe.* London: Academic Press, 150–177.

Pretty, J. N., Brett, C., Gee, D., Hine, R. E., Mason, C. F., Morison, J. I. L., Raven, H., Rayment, M. D., van der Bijl, G. (2000). An assessment of the total external costs of UK agriculture. *Agricultural Systems*, 65(2), 113–136.

Pywell, R. F., Heard, M. S., Woodcock, B. A., Hinsley, S., Ridding, L., Nowakowski, M. & Bullock, J. M. (2015). Wildlife-friendly farming increases crop yield: Evidence for ecological intensification. *Proceedings of the Royal Society B: Biological Sciences*, 282.

Ramos García, M., Guzmán, G. I. & González De Molina, M. (2018). Dynamics of organic agriculture in Andalusia: Moving toward conventionalization? *Agroecology and Sustainable Food Systems*, 42(3), 328–359.

Riley, M. (2006). Reconsidering conceptualisations of farm conservation activity: The case of conserving hay meadows. *Journal of Rural Studies*, 22(3), 337–353.

Riley, M. (2011). Turning farmers into conservationists? Progress and prospects. *Geography Compass*, 5(6), 369–389.

Riley, M. (2016). How does longer term participation in agri-environment schemes [re]shape farmers' environmental dispositions and identities? *Land Use Policy*, 52(5), 62–75.

Rosin, C. & Campbell, H. (2009). Beyond bifurcation: Examining the conventions of organic agriculture in New Zealand. *Journal of Rural Studies*, 25(1), 35–47.

Saunders, F. P. (2016). Complex shades of green: Gradually changing notions of the 'good farmer' in a Swedish context. *Sociologia Ruralis*, 56(3), 391–407.

Schewe, R. L. (2014). Letting go of 'conventionalisation': Family labour on New Zealand organic dairy farms. *Sociologia Ruralis*, 55(1), 85–105.

Scott Committee (1942). *Report of the Committee on Land Utilization in Rural Areas*. Cmd 6378. London: HM Stationery Office.

Sheail, J. (1985). *Pesticides and Nature Conservation: The British experience 1950–1975*. Oxford: Clarendon Press.

Shoard, M. (1980). *The Theft of the Countryside*. London: Temple Smith.

Siebert, R., Toogood, M. & Knierim, A. (2006). Factors affecting European farmers' participation in biodiversity policies. *Sociologia Ruralis*, 46(4), 318–340.

Silvasti, T. (2003). The cultural model of "the good farmer" and the environmental question in Finland. *Agriculture and Human Values*, 20(2), 143–150.

Smeding, F. W. & de Snoo, G. R. (2003). A concept of food-web structure in organic arable farming systems. *Landscape and Urban Planning*, 65(4), 219–236.

Strohbach, M. W., Kohler, M. L., Dauber, J. & Klimek, S. (2015). High Nature Value farming: From indication to conservation. *Ecological Indicators*, 57, 557–563.

Sutherland, L.-A. (2013). Can organic farmers be 'good farmers'? Adding the 'taste of necessity' to the conventionalization debate. *Agriculture and Human Values*, 30(3), 429–441.

Sutherland, L.-A. & Darnhofer, I. (2012). Of organic farmers and 'good farmers': Changing habitus in rural England. *Journal of Rural Studies*, 28(3), 232–240.

Świergiel, W., Pereira Querol, M., Rämert, B., Tasin, M. & Vänninen, I. (2018). Productivist or multifunctional: An activity theory approach to the development of organic farming concepts in Sweden. *Agroecology and Sustainable Food Systems*, 42(2), 210–239.

The Scottish Government (2011). *Developing High Nature Value Farming and Forestry Indicators for the Scotland Rural Development Programme*. Summary report of the Technical Working Group on High Nature Value Farming and Forestry Indicators. Edinburgh: The Scottish Government.

Tovey, H. (2002). Food, environmentalism and rural sociology: On the organic farming movement in Ireland. *Sociologia Ruralis*, 37(1), 21–37.

Vail, D. (1991). Economic and ecological crises: Transforming Swedish agricultural policy. *In:* Friedland, W., Busch, L., Buttel, F. & Rudy, A. (eds) *Towards a New Political Economy of Agriculture.* Boulder, CO: Westview Press, 256–274.

van Dijk, W. F., Lokhorst, A. M., Berendse, F. & de Snoo, G. R. (2015). Collective agri-environment schemes: How can regional environmental cooperatives enhance farmers' intentions for agri-environment schemes? *Land Use Policy*, 42, 759–766.

van Dijk, W. F. A., Lokhorst, A. M., Berendse, F. & de Snoo, G. R. (2016). Factors underlying farmers' intentions to perform unsubsidised agri-environmental measures. *Land Use Policy*, 59, 207–216.

Wang, Z. (1997). Responding to Silent Spring: Scientists, popular science communication, and environmental policy in the Kennedy years. *Science Communication*, 19(2), 141–163.

Watts, D., Little, J. & Ilbery, B. (2018). 'I am pleased to shop somewhere that is fighting the supermarkets a little bit'. A cultural political economy of alternative food networks. *Geoforum*, 91, 21–29.

Whatmore, S. (2002). *Hybrid Geographies: Natures, cultures, spaces.* London: SAGE.

Wilson, G. A. (1996). Farmer environmental attitudes and ESA participation. *Geoforum*, 27(2), 115–131.

Wilson, G. A. & Hart, K. (2000). Financial imperative or conservation concern? EU farmers' motivations for participation in voluntary agri-environmental schemes. *Environment and Planning A*, 32(12), 2161–2185.

Wilson, G. A. & Hart, K. (2001). Farmer participation in agri-environmental schemes: Towards conservation-oriented thinking? *Sociologia Ruralis*, 41(2), 254–274.

Wilson, J. D., Evans, J., Browne, S. J. & King, J. R. (1997). Territory distribution and breeding success of skylarks *Alauda arvensis* on organic and intensive farmland in southern England. *Journal of Applied Ecology*, 1462–1478.

Winter, M. (2005). Geographies of food: Agro-food geographies – food, nature, farmers and agency. *Progress in Human Geography*, 29(5), 609–617.

Winter, M. (2013). Environmental issues in agriculture: Farming systems and ecosystem services *In:* Murcott, A., Belasco, W. & Jackson, P. (eds) *The Handbook of Food Research.* London: Bloomsbury Academic, 192–208.

Worster, D. (1990). Transformations of the Earth: Toward an agroecological perspective in history. *The Journal of American History*, 76(4), 1087–1106.

9 Can farmers deliver?

Prospects in an era of food 'challenge'

Introduction

In this final chapter, we turn our eye to the future and seek to assess the prospects for the emergence of a holistic, socially sensitive sustainable agriculture in the UK. We first consider first, to the current policy debate around Brexit as this now provides the context in which much of agriculture's future will be played out. We then look at the challenges and prospects for growing more food, achieving food security, and then similarly the prospects for environmental security within an agricultural context. In both instances, we point to the importance of the neglected social dimension – understanding and solving the social question is a pre-requisite for the UK to achieve food and environmental security, particularly if we are to avoid 'exporting'.

Brexit: an (unexpected) twist in the story

When we first conceived of this book and commenced some of its writing, we expected a substantial emphasis on the Common Agricultural Policy and, in particular, the prospects for further incremental reform to the CAP, a subject with which we have long been engaged (for example, Lobley and Butler 2010 and Winter et al. 1998). At the time of writing in the early spring of 2018, it is impossible to say what the UK's trading relationship with the European Union will be after Brexit, even assuming that the UK does in fact leave the EU. All we can do at this point is give an indication of the policy thinking as it is currently available to us and do so, in particular, in the context of the social sustainability issues at the heart of this book. Things may change!

The extraordinary range of possible trade outcomes is well illustrated in a modelling exercise by Feng et al. (2017). Taking three possible trade scenarios (Bespoke Free Trade Agreement with the EU, WTO Default, and Unilateral Trade Liberalisation), some illustrative outputs are shown in Table 9.1.

But these data, and others like them, do not take into account the wider context of how agricultural policy is likely to be reframed in a post-Brexit world, assuming that is that the eventual Brexit agreement does not tie the UK back into CAP through a Norwegian style relationship with the EU. The

Table 9.1 Percentage changes in UK commodity prices, under three trade scenarios, 2019–2025

	Bespoke FTA	*WTO*	*UTL*
Beef	+3	+17	−45
Sheep	−1	−30	−29
Milk and dairy	+1	+30	−10
Pigs	0	+18	−12
Wheat	−1	−4	−5

Source: Feng et al., 2017, p.28

government`s agricultural consultation paper on the future of agriculture, *Health and Harmony*, was published in the spring of 2018 (Defra 2018). Some of the key *Health and Harmony* policy objectives are as follows:

- A dynamic, more self-reliant agriculture industry;
- A reformed agricultural and land management policy to deliver a better and richer environment in England;
- Public money for public goods, such as environmental enhancement and protection, better animal and plant health, animal welfare, improved public access, rural resilience and productivity;
- An implementation period and an 'agricultural transition' period to give farmers time to prepare for new trading relationships and an environmental land management system;
- Direct payments to continue during the 'agricultural transition' but with reductions to fund pilots of environmental land management schemes;
- An enabling environment for farmers to improve their productivity and add value to their products, so they can become more profitable and competitive, including a proposal to further reduce and phase out direct payments in England completely by the end of the 'agricultural transition' period, which will last a number of years beyond the implementation period;
- Simplification of Countryside Stewardship schemes, cross-compliance and greening requirements before moving to a new regulatory regime.

No one disputes that the UK agricultural sector faces a highly uncertain future as a result of Brexit and farmers will need to be supported through the transition from a high dependency on direct payments under the CAP. But precisely how this might happen and the manner in which public goods are identified and defined is tantalisingly unclear. Moreover, narratives of productivity and self-reliance sit uneasily alongside narratives of public good provision and regulation. And there are issues that were inadequately addressed in *Health and Harmony*. It is gratifying to read on page 12 and again on page 34 that 'the UK's farmed land is rich in social and cultural heritage and significance'

(Defra 2018 p.12). But the empirical realities, causal relationships and policy implications are not developed. Nor, despite the title, are the links between health and food made explicit, still less are they seen as an important candidate for public policy intervention. Defra do not seem to have heeded the call for health to be central to farming policy made by Tim Lang et al. (2017) well in advance of the publication of *Health and Harmony*:

> Our main concern is that civil society, academics and external voices – whatever their specialisms – should unite around the call for the new Food Brexit Framework to locate food as a central (and cross-departmental) part of UK public policy in progressing and creating a more resilient, robust food system in the UK. This should be one which is capable of delivering sustainable and future generational diets, healthy lifestyles and environments for its increasing and diverse population. … In order to achieve these goals the UK will need a statutory framework which creates and promotes a unique and novel UK approach to One Nation Food. The new UK statutory framework will need cross-departmental and devolved authority support and commitment and not just be associated with Defra or any other single department. It should include the creation of a Standing Committee or Commission on Food and Agricultural Policy.
>
> (Lang et al. 2017, p.76)

An underlying and often unspoken issue in all the debate about public goods and forging a new distinctive policy path for UK agriculture is international trade. It is disconcerting to say the least that, at the time of writing, international trade wars appear to be breaking out, driven by the politics of isolationism and nationalism associated with the Trump presidency in the USA. It is perhaps equally disconcerting that, notwithstanding this recent flurry of protectionism, the neo-liberal project of global trade is so deeply rooted that one commentator refers to the 'trade-ification of the food sustainability agenda' (Clapp 2017), arguing that international trade has become so normalised as a means of delivering food system sustainability based on the efficiency gains from trade, that alternative approaches, such as food sovereignty, are squeezed out. Food sovereignty at its core, 'is a set of goals comprised of strengthening community, livelihoods and social and environmental sustainability in the production, consumption and distribution of nutritious and culturally appropriate food' (Desmarais and Wittman 2014, p.1155). On the face of it, these seem to be an unexceptionable set of goals and values but they are antithetical to those fully committed to global trade and free markets. The association of food sovereignty with La Via Campesina – a radical international peasants' movement arguing for local food under the control of local producers – has bolstered this view and in the UK has been associated with national trade protectionism and self-sufficiency (e.g. Foresight 2011), although this is only a partial understanding of food sovereignty (Spencer et al. 2014). For Foresight (2011), food security was a more acceptable term than food sovereignty and given the association of food sovereignty with opposition to trade maybe, pragmatically, framing agriculture's role in those terms is a more plausible path, especially if 'food security' can be seen as an aspect of 'public good'. Food does help embed economic argumentation, whether of private market goods or public market-failure goods, into a deeper moral and cultural undertaking. The framing of 'food security' as an emerging imperative is consistent with the idea held by some farmers and scientists that 'the balance has been tipped too recklessly towards environmental sustainability and away from food production' (Lowe et al. 2009, p.26).

Tackling the social challenge and building resilience

In a recent paper, Calus and van Huylenbroeck (2010) remind us that the key to the success of the family farm is the use of labour, largely the labour provided by the farmer and members of the farm family. We would go further and say that the key to the success of the family farm is people. In the Foreword to his Report on the Future of Farming, David Fursdon writes that 'When someone visits a farm they will look for what is important to them. Everyone will look for different things. I am normally interested in the people. It is largely up to them whether or not it succeeds' (Fursdon, 2013).

One of our objectives in writing this book has been to begin to address the neglected social dimension of UK farming. Our use of the term 'social' is deliberately broad encompassing what economists would term 'human capital' (i.e. knowledge and skill), human relationships (e.g. between tenant and landowner, producer and consumer, farmers and non-farmers) and personal well-being. We argue that the personal *is* social. Low levels of well-being, poor self-esteem and depression are societal problems and stem, at least in part, from what many farmers perceive to be a fracturing of society, creating an 'us and them' situation where farmers can all too easily feel isolated, on the outside, misunderstood and unwanted. Add to this low and fluctuating incomes; dependency on public payments; asymmetrical power relations in the supply chain; a difficult and sometimes dangerous working environment; and now a potential radical change in the policy environment and it is clear that farming households face a tough challenge.

Given that the challenge is multifaceted, so must be the response. A good place to start to build resilience is to address the issue of farmers' place in society and their relationship with non-farmers. Farmers may produce the raw materials for the nation's largest manufacturing sector (Food & Drink) but they are in a minority in all but the most remote rural areas. The combination of rural social change, broader societal trends (particularly around our relationship with food and nature) and changes within farming itself (e.g. increased mechanisation) have all served to distance farmers from other members of the community, resulting in a group of people who produce our food and manage our environment who are quite culturally distinct but also frequently socially isolated. Initiatives such as Open Farm Sunday,[1] in which hundreds of farmers open their farms to the public, can help build bridges and provide mutually educative experiences. But what about those farmers who, for a variety of reasons, do not want to or are unable to take part in such activities? Previous research (Milbourne at al. 2001) suggested that one approach to reduce the potential for conflict between farmers and non-farming residents of rural areas and to build understanding about farming would be to produce local area profiles to 'provide information about the local land-based economy in the context of the wider social economy of the area' (Milbourne at al. 2001, p.139). Obviously, such profiles would provide important and useful information but, where it exists, division between farming and non-farming members of communities does not arise from a simple information deficit. Farmers can often be heard to say that 'the public should be educated about where their food comes from'. Education and information obviously has a role but it is the building of relationships between farmers and non-farmers that is required. The process of compiling local area profiles, if done in a collaborative manner, could help with this.

Building relationships, developing mutual understandings, and providing social support networks are all vital in improving the well-being of farmers and tackling the problem of social isolation. However, we must also be alert to the very real issue of depression and mental ill-health in the farming community. This is clearly an area requiring further interdisciplinary research but, on a practical level, policymakers need to be aware of the potential for Brexit and a radically different British Agricultural Policy to threaten the survival of many family farms should they lose significant public support payments. In writing about his decision to sell the family farm, Kuehne (2013, p. 211) says that

> even though I feel as though I was more prepared to leave farming than others I still experienced feelings of uncertainty, failure and loss of relevance (I wasn't important to the farm or the land any more). I experienced a loss of identity, and a loss of the socio-cultural rewards gained from being a farmer. In addition to this I also felt that I had abandoned an obligation that I felt mostly to my family but also to my friends and colleagues to continue farming.

For those who are unprepared, the prospect of leaving the farm could have a profound impact on well-being and mental health. Even in the absence of Brexit, it is time that the well-being and mental health of the farming community is addressed.

Beyond issues of well-being, there are several actions that can be taken to enhance the resilience of farming in Brexit UK. Wilson's (2014) analysis of the characteristics of 'improved' farming businesses identified a number of common characteristics, notably the wide range of different life-skills and aptitudes that are required. A successful farmer has to have business acumen in terms of financial management (attention to costs and margins) as well as technical knowledge and know-how (agronomy, husbandry, mechanical skills), market knowledge, and social/emotional/familial intelligence and awareness. If all this seems to be a tall order for a hard-pressed farming sector, we need to remember that these are the attributes to run only a successful farm business. McElwee (2008) has produced a taxonomy of farmers that delineates the ways that farmers can be viewed as entrepreneurial in the context of the various avenues of non-farming diversification they may also pursue. He distinguishes between four entrepreneurial paths a farmer may follow, highlighting different ways of being 'a farmer' and the possibilities and constraints on entrepreneurship in all cases:

- Farmer as farmer (engaged in traditional land-based economic activity);
- Farmer as entrepreneur (innovative, opportunity-oriented combined with changing, flexible and diverse economic activities);
- Farmer as contractor (owning specific skills/expertise and experience coupled with possible ownership of 'plant');
- Rural entrepreneur (ownership of farm, land or business).

There is clearly, therefore, a need to help farmers (and aspiring farmers) to develop good management and technical skills to assist with the effective day-to-day management of a successful farm business. As well as through formal training and education, the sector could benefit from adopting life-long learning through regularly accessing advice, support and information to help inform business decisions. The upskilling of agriculture implied by this could be supported by the much-mooted idea of a professional accreditation. The Institute for Agricultural Management has been promoting the idea of the Chartered Agriculturalist (C.Agric). While not necessarily a 'licence to farm', conferral of C.Agric would be formal recognition of a level of professional competency, skills and knowledge against a set of criteria. Chartered Agriculturalists would also be required to demonstrate continuing improvement in skills, knowledge and education to maintain the title.

The upskilling and further professionalisation required to help build greater resilience in the sector cannot occur in a vacuum. Rather, it should be seen as part of a process that drives a knowledge-rich agriculture and this requires access to advice. Since the demise in the 1980s of a national advisory service (ADAS) free to farmers at the point of delivery, there have been numerous bespoke initiatives to provide advice to farmers. The so-called Agricultural Knowledge and Information System or AKIS (Prager and Thomson, 2014) is different in and within each of the four countries of the UK and, taken as a whole, is characterised by its ad hoc and diffuse nature. There are separate initiatives for conservation advice and for pollution advice, usually publicly funded; agronomy and technology advice usually comes from the private sector. There have been publicly funded initiatives around farm business management advice and numerous consultants offer similar services. Reports and papers lamenting the lack of coordination and inconsistences abound and have done for many years (e.g. Winter, 1995). The kind of integrated and wide-ranging advice and business support that is needed to address performance issues is not readily available. Coordination and targeting is vital. Succession planning may be more important than agronomy or vice versa depending on circumstances. The ability to determine this and act accordingly requires advisors and other knowledge brokers to abandon their professional silos. While in the current financial climate there is little prospect of fresh public sector funding to address this problem, and certainly not to re-create a national advisory service, there is more that could be done to address the need for coordination. This almost certainly needs government encouragement and Brexit provides an opportunity for this.

New blood and generational change

Building the resilience of agriculture to survive in a post-Brexit world requires that attention is given to the next generation of farmers as well as existing farmers. With land prices high and relatively limited amounts of land appearing on the market for sale or rent each year (see Chapter 3), the tendency for existing larger farms, or even large non-farming interests, to buy or rent land when it comes available means there are few opportunities for new entrants. This is not an issue confined to the UK. The European Economic and Social Committee identified land grabbing and land concentration as a threat to family farms (EESC, 2015). To a country as steeped in free-market principles as the UK, some of their conclusions are both striking and challenging:

> Land is no ordinary commodity which can simply be manufactured in larger quantities. Given that the supply of land is finite, the usual market rules should not apply. Ownership of land and land use must be subject to greater regulation. In view of the distortions that have been observed, the EESC considers it necessary to develop a clear model for agricultural structures at both Member State and EU level, which will have implications for land use and land rights.
>
> (EESC, 2015, p.3)

Some countries, such as Denmark, France, Switzerland and Sweden, have longstanding regulations on who can occupy agricultural land which are designed to maintain family farming and the social fabric of rural communities. In France, land transactions are monitored by regional land authorities (Sociétés d'Aménagement Foncier et d'Etablissement Rural, (SAFER)), charged with supporting farmers – especially young farmers – and ensuring transparent agricultural land markets. The UK has no such regulations in force (although the land reforms proposed by the Scottish government might be considered by some as a step away from an unfettered free market in land). In that context, and with no expectation of land market regulation, we need to look at private initiatives to encourage responsible land transactions and reforms to the tax regime. Rural estates can play an important role here, for example through the creation of opportunities for new farm businesses by investing in the provision of new housing for existing tenants in order to facilitate new entrants.

There is limited point, however, in seeking to assist new entrants without also addressing the interconnected issue of an ageing farm population and frequent lack of succession planning. The long-term trend of an ageing farm population is common among many OECD countries and is caused by a low rate of exit and equally low rate of entry. A combination of a deep personal commitment to agriculture and the operation of the tax system can conspire to make retirement an unattractive option for many farmers (Winter and Lobley, 2016). Nevertheless, a thriving agricultural sector equipped to face the challenges outlined by Wilson (2014) and survive in a post-Brexit world requires renewal of the family farm system via familial intergenerational succession where an appropriate family successor (or successors) is available. This is an issue much talked about in agricultural circles and, although recent years have seen an increase in awareness, research by *Farmers' Weekly* indicated that 60 per cent of family farms do not have a succession plan. Succession planning is a 'slow burn' and it is important that high levels of awareness are maintained and that farming families are encouraged to progress from being aware of the issues to being facilitated to develop a succession road map and then engage with professional services providers (e.g. accountants and solicitors) in order to develop an appropriate plan.

Where a familial successor is not available, or simply where a landowner wishes to offer an opportunity to a 'new entrant', then a matchmaking service such as that recently launched by the Fresh Start Land Enterprise Centre has much to offer. Such initiatives have long existed in the United States. The Fresh Start 'Land Partnerships Service' can help facilitate a variety of arrangements, including contract farming, licences and profit of pasturage, share farming, partnerships, conventional tenancies and long-term lets. Ingram and Kirwan's (2011, p.917) analysis of a previous matchmaking initiative in Cornwall points to 'a deep rooted reluctance amongst participants in the initiative to enter formal long-term joint ventures due to differing motivations, expectations, and concerns about their respective responsibilities in the working relationship and about the validity of the legal framework'. The new initiative builds on this experience (and that from elsewhere) and has industry backing. However, it must be recognised that matchmaking (and the all-important mentoring service also offered by Fresh Start) and succession planning require personal interaction over an extended period of time and need to be backed by long-term funding in recognition of the time required to build relationships and gain the trust that will be so necessary for successful outcomes. It is likely that incoming, would-be land-based entrepreneurs would be in a stronger position if they were also able to undertake the sort of professional certification process discussed above.

Farming families should be encouraged to see succession planning as an investment in the future of their business and family but, like any investment, they will value some contribution towards the cost. In addition to such incentives, consideration could be given to making business loans conditional on succession planning (where appropriate). Succession planning services should be encouraged. However, it is vital that anyone engaged in offering succession planning advice understands that successful succession involves much more than the transfer of tangible assets. The transfer of intangible assets and delegation of managerial control are essential for successful succession, as is retirement planning. Succession is not simply a tax accounting and legal issue. It can involve complex psychology; changing roles within the business, family and community, and can test interpersonal relationships. As such, it requires advisors and facilitators (a multi-disciplinary support team) who are aware of these issues and who can help families steer a clear course through the succession process.

In instances where farming families do not have a familial successor but there is a desire for the farm to continue to be owned by the family, share farming agreements can offer an entry route for undercapitalised new entrants. The Country Land and Business Association (CLA) argue that 'the adoption of share farming as a farm business structure is eminently suited to many of the challenges the industry currently faces' (CLA, 2014, p.5). The CLA have clearly set out the benefits of such arrangements to landowners and operators. Share farming is not suitable for everyone but we agree with the CLA that it can play an important role in bringing new blood into agriculture and endorse the recommendation of the Future of Farming Review (Fursdon, 2013 p.3) that 'joint equity and flexible business sharing schemes should be encouraged to enable a gradual handover of businesses, particularly where there is no natural successor'.

In addition, there is a need to look carefully at the obstacles to change, which may include the difficulty of securing a retirement property. In some instances initiatives such as the Strategic Rural Housing Scheme operated by the Addington Fund can help find a new home for existing farmers and their families. In addition, we would argue that consideration should be given in planning policy to allow farmers of retirement age to build a retirement house when they agree to facilitate new entrants through farm business tenancies (FBTs), share farming or land purchase. A similar proposal has also been put forward by the CLA (CLA 2018).

Another route for new entrants is provided by FBTs, although an evaluation by the University of Plymouth (Whitehead et al, 2002) suggested that new entrants felt unable to compete with established businesses in bidding for FBTs. The apparent short-term nature of many FBTs has also been criticised; the Tenant Farmers' Association is actively campaigning for the length of FBTs to be at least 10 years, suggesting a number of ways in which this might be facilitated such as including tax incentives and disincentives. Longer-term FBTs would certainly provide additional security and more of an incentive to develop a business and offer routes to progression.

Generally, we strongly endorse the recommendation from the *Future of Farming Review* that

> low numbers of retiring farmers can restrict opportunities and this is exacerbated by the CAP and the inheritance tax framework. We need coordinated action to help support farmers to plan for retirement and succession on both their own and on rented farms at an earlier age and modify the detailed application of Agricultural Property Relief which encourages them to farm until death.
>
> (Fursdon, 2013, p.3)

We would also stress the importance of facilitating a dignified withdrawal for elderly farmers and one which recognises the value of their contribution, knowledge and skills. Brexit may well take care of the problem of the CAP, or more precisely CAP support payments, acting as a disincentive to retirement, but the role of capital tax reliefs associated with agricultural land and businesses should be examined.

The supply chain

As discussed in Chapter 4, farmers' relative weakness in the supply chain is a fundamental and continuing problem. Shortening the food chain through direct retail, including participation in alternative food networks, clearly offers a pathway for some farmers. Organic, local, high-quality, premium specialty foods – all differentiated from the offerings of mainstream food manufacturers and retailers – have generated huge academic interest (e.g. Goodman et al., 2014) but have touched a relatively small proportion of farmers. There is considerable scope for more development in this area, which also has the added benefit of helping to build relationships between producers and consumers.

Another way in which family farmers can strengthen their position is through collaboration and cooperation, for both buying and selling. This can help achieve some of the benefits of scale and enable farmers to retain a greater proportion of the end price for their produce. Recent evidence from the Defra-funded Sustainable Intensification Research Platform indicates that cooperation among farmers in a variety of forms is higher than is often thought, challenging 'the stereotype often perpetuated by farmers themselves … that British farmers are staunchly independent actors who are disinterested in collaborating' (Morris et al, 2016, p.135).

The benefits of shortened supply chains have by-passed many family farmers. There are two reasons for this. First, the necessary investment of capital and time have often been beyond the means of the farmers with limited resources. Second, the policy focus in developing these has often been on consumers and on wider environmental issues focusing on issues such as health, food miles, and organic or ecological agriculture. There is a need to give greater emphasis to farmers within local food movements. Rural development leaders need to consider how best to engage and empower family farms farmers within local food initiatives.

Conclusion: towards resilient family farms in Brexit UK

At the time of writing we stand on the threshold of an unknown policy future for agriculture. Despite many decades of support, it seems clear that the model of direct income payments for farmers will not continue in its current form. Some level of financial support to farmers is likely to be ongoing in return for the provision of 'public goods' and services, but it is not known if this will equate to a similar level of support to a similar number of recipients as currently exists. However, what is known is that whatever the future direction for agricultural policy, for many farmers commercial objectives and pressures are likely to remain at the forefront of farming as an occupation. The ability of farmers to respond to the changing market and policy incentives while achieving income objectives will remain paramount.

Family farming has proved remarkably resilient. It has done so in part because of that combination of family and enterprise that characterises all family businesses. Like other family businesses, farming is characterised by a marked degree of emotional ownership, to the extent that farmer and farm occupy a shared space and shared identity. This is not easily given up. However, it is important to recognise that there is not necessarily a future for all family farms. As we have seen, powerful economic forces are driving changes in farm size structures and there is a limit to which they can be resisted in the absence of fundamental change in global economic systems. This should not necessarily be seen as a problem. The term 'family farm' obscures the heterogeneity exposed in various parts of the book. The term 'family farm' is really just a shorthand way of describing a spectrum of potentially very different farming, business and family situations, ranging from retirement holdings, lifestyle farms, part-time farms, main living farms and so on. This heterogeneity in the sector is likely to be reflected in a range of different futures for different farms. As we have seen, there are many different ways of being a farmer.

Note

1 https://farmsunday.org/.

References

Calus, M. & van Huylenbroeck, G. (2010). The persistence of family farming: A review of explanatory socio-economic and historical factors, *Journal of Comparative Family Studies*, 41(5), 639–660.

CLA (2014). *An Option for Enterprising Farmers: CLA's assessment of share farming*. London: CLA.

CLA (2018). *Policy Briefing (England) 1: Homes for Retiring Farmers*. London: CLA.

Clapp, J. (2017). The trade-ification of the food sustainability agenda, *Journal of Peasant Studies*, 44(2), 335–353.

Defra (2018). *Health and Harmony: The future for food, farming and the environment in a Green Brexit*, Cm 9577. London: HMSO.

Desmarais, A. A. & Wittman, H. (2014). Farmers, foodies and First Nations: Getting to food sovereignty in Canada, *Journal of Peasant Studies*, 41(6), 1153–1173.

EESC (2015). *Land Grabbing – A Warning for Europe and a Threat to Family Farming*. Opinion of the European Economic and Social Committee. NAT/632. Brussels.

Feng, S., Patton, M., Binfield, J. & Davis, J. (2017). 'Deal' or 'No Deal'? Impacts of alternative post-Brexit trade agreements on UK agriculture, *EuroChoices*, 16 (3), 27–33.

Foresight. (2011). *The Future of Food and Farming*. Final Project Report. London: The Government Office for Science.

Fursdon, D. (2013) *Future of Farming Review Report*. London: Defra.

Goodman, D., Dupuis, E. M. & Goodman, M. K. (eds) (2014) *Alternative Food Networks*. London: Routledge.

Ingram, J. & Kirwan, J. (2011). Matching new entrants and retiring farmers through farm joint ventures: insights from the Fresh Start Initiative in Cornwall, UK, *Land Use Policy*, 28(4), 917–927.

Kuehne, G. (2013). My decision to sell the family farm, *Agriculture and Human Values*, 30(2), 203–213.

Lang, T. Millstone, E. & Marsden, T. (2017). *A Food Brexit: Time to get real*. Brighton: University of Sussex, Science Policy Research Unit.

Lobley, M. & Butler, A. (2010). The impact of CAP reform on farmers' plans for the future: some evidence from South West England, *Food Policy*, 35(4), 341–348.

Lowe, P. Woods, A. Liddon, A. & Phillipson, J. (2009). Strategic land use for ecosystem services, In Winter, M. & Lobley, M, eds. *What is Land For? The Food, Fuel and Climate Change Debate*. London: Earthscan. pp. 23–46.

McElwee, G. (2008). A taxonomy of entrepreneurial farmers, *International Journal of Entrepreneurship and Small Business*, 6(3), 465–478.

Milbourne, P. Mitra, B. & Winter, M. (2001). *Agriculture and Rural Society: Complementarities and Conflicts between Farming and Incomers to the Countryside in England and Wales*. Cheltenham: Countryside and Community Research Unit.

Morris, C., Jarratt, S., Lobley, M. & Wheeler, R. (2016). *Baseline Farm Survey – Final Report*. Report for Defra project LM0302 Sustainable Intensification Research Platform Project 2: Opportunities and Risks for Farming and the Environment at Landscape Scales.

Prager, K. & Thomson, K. (2014). *AKIS and Advisory Services in the United Kingdom*. Report for the AKIS inventory (WP3) of the PRO AKIS Project. Aberdeen: The James Hutton Institute.

Spencer, A., Morris, C. & Seymour, S. (2014). *Food sovereignty in the UK* [Online]. Available at: https://blogs.nottingham.ac.uk/globalfoodsecurity/2014/01/10/food-sovereignty-in-the-uk/ [accessed 17/05/18].

Whitehead, I., Errington, A., Millard, N. & Felton, T. (2002). *An Economic Evaluation of the Agricultural Tenancies Act 1995*. Final Report for DEFRA. London: Defra.

Wilson, P. (2014). Farmer characteristics associated with improved and high farm business performance, *International Journal of Agricultural Management*, 3(4), 191–199.

Winter, M. (1995). *Networks of Knowledge: A review of environmental advice, training, education and research for the agricultural community in the UK*. Report to WWF.

Winter, M. Gaskell, P. Gasson, R. & Short, C. (1998). *The Effects of the 1992 Reform of the Common Agricultural Policy on the Countryside of Great Britain*. Cheltenham: Countryside and Community Press and Countryside Commission.

Winter, M. & Lobley, M. (2016). *Is there a Future for the Small Family Farm in the UK?* Report to the Prince's Countryside Fund. London: Prince's Countryside Fund.

Index

Note: Page numbers in **bold** refer to tables and in *italics* to figures.

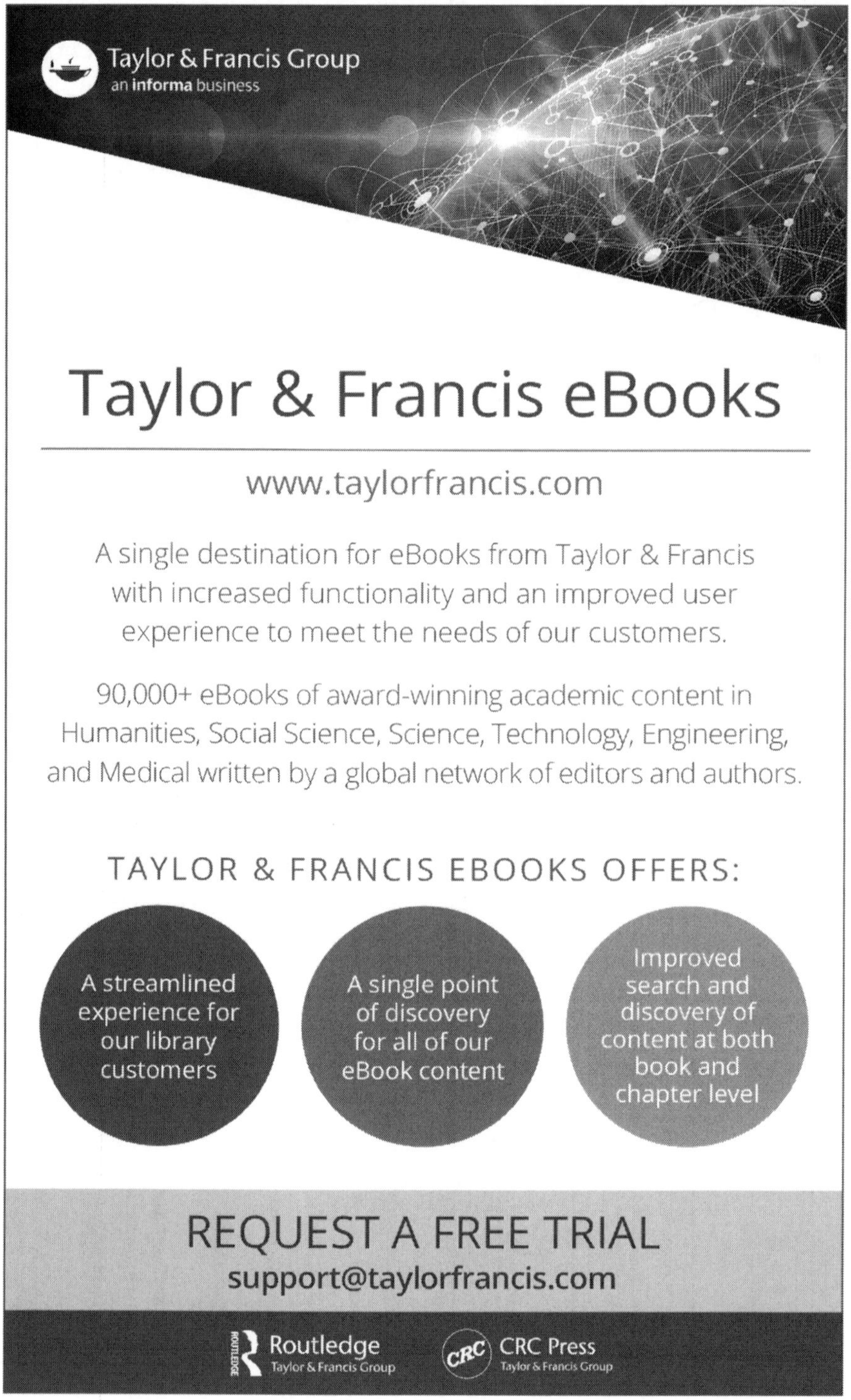